半导体与集成电路关键技术丛书

垂直型 GaN 和 SiC 功率器件
Vertical GaN and SiC Power Devices

[日] 望月和浩（Kazuhiro Mochizuki） 著

黄　锋	段宝兴	柏　松	万成安	
赵小宁	刘凌旗	王传声	张　明	
张志国	王英民	王雨茗	石　杨	
魏敬和	李　静	何　君	赵元英	译
赵金霞	史　超	王淑华	唐林江	
飞景明	刘斯扬	邓小川	孙伟锋	
孙叔翔	张　茹	谭　骥	李　赟	
费晨曦				

杨银堂　李　晨　审校

机械工业出版社

近年来，以氮化镓（GaN）和碳化硅（SiC）等宽禁带半导体化合物为代表的第三代半导体材料引发全球瞩目。第三代半导体广泛应用于新一代移动通信、新能源汽车、物联网和国防电子等产业，已成为国际半导体领域的重点研究方向。

本书主要介绍垂直型 GaN 和 SiC 功率器件的材料、工艺、特性和可靠性等相关技术，内容涵盖垂直型和横向功率半导体器件的比较，GaN 和 SiC 的物理性质、外延生长、制备工艺、主要器件结构与特性，以及垂直型 GaN 和 SiC 功率器件的可靠性研究等。

本书适合从事 GaN 和 SiC 功率半导体技术的科研工作者、工程师阅读，也可作为高等院校微电子科学与工程、电力电子技术等相关专业的教材。

Vertical GaN and SiC Power Devices, by Kazuhiro Mochizuki, ISBN：978-1-63081-427-4.
@ 2018 ARTECH HOUSE

This title is published in China by China Machine Press with license from Artech House. This edition is authorized for sale in the Chinese mainland, excluding Hong Kong SAR, Macao SAR and Taiwan. Unauthorized export of this edition is a violation of the Copyright Act. Violation of this Law is subject to Civil and Criminal Penalties.

本书由 Artech House 授权机械工业出版社在中国大陆地区（不包括香港、澳门特别行政区及台湾地区）出版与发行。未经许可之出口，视为违反著作权法，将受法律之制裁。

北京市版权局著作权合同登记 图字：01-2019-0964 号。

图书在版编目（CIP）数据

垂直型 GaN 和 SiC 功率器件/（日）望月和浩著；黄锋等译.
—北京：机械工业出版社，2022.5（2023.12 重印）
（半导体与集成电路关键技术丛书）
书名原文：Vertical GaN and SiC Power Devices
ISBN 978-7-111-70502-4

Ⅰ.①垂… Ⅱ.①望…②黄… Ⅲ.①功率半导体器件 Ⅳ.①TN303

中国版本图书馆 CIP 数据核字（2022）第 056489 号

机械工业出版社（北京市百万庄大街 22 号　邮政编码 100037）
策划编辑：付承桂　　　责任编辑：付承桂　杨　琼
责任校对：陈　越　张　薇　封面设计：马精明
责任印制：单爱军
北京虎彩文化传播有限公司印刷
2023 年 12 月第 1 版第 2 次印刷
169mm×239mm・14.5 印张・240 千字
标准书号：ISBN 978-7-111-70502-4
定价：99.00 元

电话服务　　　　　　　　　网络服务
客服电话：010-88361066　　机 工 官 网：www.cmpbook.com
　　　　　010-88379833　　机 工 官 博：weibo.com/cmp1952
　　　　　010-68326294　　金 书 网：www.golden-book.com
封底无防伪标均为盗版　机工教育服务网：www.cmpedu.com

译者的话

近年来，以氮化镓（GaN）和碳化硅（SiC）等宽禁带半导体为代表的第三代半导体材料和器件的发展引起了全球科技界和工业界的广泛关注。与第一代的硅（Si）和第二代的砷化镓（GaAs）等相比，第三代半导体材料由于禁带宽度大、击穿场强高、热导率高、电子饱和速度快以及载流子迁移率高，使得器件导通电阻减小，承载能量密度更高，可以在200℃以上高温下工作，提升了器件的整体能效和可靠性，可广泛应用于新一代移动通信、新能源汽车、物联网和国防电子等产业，已成为国际半导体领域的重点研究方向。

本书是介绍垂直型GaN和SiC功率器件理论及相关技术的专著，内容涵盖了垂直型和横向功率半导体器件的比较、GaN和SiC的物理性质、材料生长、外延和关键制作工艺、主要器件结构与特性，以及垂直型GaN和SiC功率器件的可靠性研究等，选题新颖，内容先进，指导性强。在当前世界各国竞相研究发展第三代半导体技术的背景下，本书的出版非常重要和及时，必将为国内相关专业的技术人员和高校师生提供重要的技术支撑和教学参考。

从本书的策划、翻译到最后完稿付梓，机械工业出版社给予了大力支持和帮助。西安电子科技大学杨银堂教授和中国电子科技集团李晨研究员对本书进行了认真审校，在此特别表示感谢。由于本书专业性强，虽经译者反复多轮译校，但囿于译者的水平和时间，疏漏之处在所难免，望读者不吝指正。

译者

原书前言

与硅材料相比,氮化镓(GaN)和碳化硅(SiC)材料具有优异的物理特性,如临界击穿电场强度比硅材料高 10 倍以上,热导率高 33%~330%。因此,它们有望成为高效电力电子系统中功率器件的最优材料。以 GaN 为例,近年来,高质量 GaN 衬底的进展极大地提高了垂直型功率器件的性能。但是在最新出版的一本关于 GaN 和 SiC 功率器件的书籍中[1],只用 10 页的篇幅对这种性能的改善进行了简要描述。鉴于这种情况,本书旨在为从事 GaN 和 SiC 晶体生长、加工和功率半导体器件设计领域的学生、研究人员和工程师提供垂直型 GaN 和 SiC 功率器件的分析和比较。

在 2016 年国际化合物半导体研讨会上,我受邀做了题为"垂直型 GaN 双极器件:从光子回收中获得竞争优势"[2]的演讲,从那之后我开始写这本书。因此,这本书的重点部分是光子回收(见第 4 章),这不仅对提高有效少数载流子寿命很重要,而且对提高深受主的电离率也很重要。GaN 中的光子回收可归因于非常大的边缘电流,而电流则流过非自对准台面型 p-n 结(见第 3 章)。由于对流过 p-n 结的正向电流只进行了一维处理,因此光子回收这一现象在任何书籍中都没有描述。

此外,我所知的关于 GaN 和 SiC 功率器件的书籍均没有充分描述肖特基结,也就是高电场强度下电子迁移率变得相当低,但还是把热离子发射描述为电流传输的限制过程。如第 8 章所述,其实扩散过程更为重要,尤其是当离子注入引起的损伤或高温下的声子散射降低了电子迁移率。

对于单极功率开关器件,当击穿电压为 6.5kV 时,SiC 超结功率开关器件具有最低的导通电阻。尽管沟槽填充外延是制备 Si 和 SiC 超结器件必不可少的方法,但沟槽填充工艺对于 Si 和 SiC 来说是完全不同的。第 6 章解释了这种外延生长工艺。

以前关于功率半导体器件的书中没有涉及的其他主题包括:基于报道结果的垂直型和横向功率半导体器件的比较(见第 1 章),SiC 中铝和硼的分布模型

（见第 7 章），GaN 双极晶体管（见第 10 章），垂直型 GaN 和 SiC 功率器件的结终端和可靠性比较（见第 11 章和第 12 章）。

这本书的写作得到了许多专家的帮助和有价值的讨论，包括法政大学的三岛友（Tomoyoshi Mishima）教授的外部光子回收（见第 4 章）；东京大学荣誉退休教授西野太郎（Tatau Nishinaga）的螺旋增长过饱和度和吉布斯-汤姆逊效应（见第 6 章）；日本高级科学技术研究所铃木俊二（Toshikazu Suzuki）教授的肖特基结（见第 8 章）；还有阿泰克（Artech）出版社的匿名审稿人对全文的审稿。我还要感谢日立公司的岛本康弘（Yasuhiro Shimamoto）博士、石明孝（Akio Shima）博士和森喜佑（Yuki Mori）博士在修改手稿方面给予的支持。

望月和浩（Kazuhiro Mochizuki）

参 考 文 献

[1] Baliga, B. J., *Gallium Nitride and Silicon Carbide Power Devices,* Singapore: World Scientific, 2017, pp. 391–400.

[2] Mochizuki, K., "Vertical GaN Bipolar Devices: Gaining Competitive Advantage from Photon Recycling," *International Symposium on Compound Semiconductors*, Toyama, Japan, June 26–30, 2016, paper ThB2-1.

目录

译者的话
原书前言

第1章　垂直型与横向功率半导体器件 ················· 1
  1.1　引言 ·· 1
  1.2　典型功率半导体器件特性 ····················· 3
  1.3　垂直型与横向单极功率半导体器件 ············· 4
    1.3.1　垂直型和横向单极功率开关器件 ·········· 5
    1.3.2　垂直型和横向单极功率二极管 ············ 8
  1.4　总结 ······································ 10
  参考文献 ······································ 11

第2章　GaN 和 SiC 的物理性质 ················· 13
  2.1　引言 ······································ 13
  2.2　晶体结构 ·································· 14
    2.2.1　AlN 和 GaN 的晶体结构 ················ 14
    2.2.2　SiC 的晶体结构 ······················· 17
    2.2.3　晶体缺陷 ····························· 18
  2.3　能带 ······································ 20
  2.4　杂质掺杂 ·································· 22
    2.4.1　n 型掺杂 ····························· 23
    2.4.2　p 型掺杂 ····························· 24
  2.5　载流子迁移率 ······························ 25
  2.6　碰撞电离 ·································· 26
  2.7　品质因数 ·································· 27
  2.8　总结 ······································ 29

参考文献 ··· 29

第3章　p-n 结 ·· **34**

3.1　引言 ·· 34
3.2　扩散 ·· 34
3.3　连续性方程 ··· 35
3.4　载流子复合寿命 ··· 36
　　3.4.1　带间复合寿命 ·· 36
　　3.4.2　间接复合寿命 ·· 38
　　3.4.3　俄歇复合寿命 ·· 39
　　3.4.4　载流子复合寿命的整体表达式 ··· 39
3.5　一维 p^+n 突变结的耗尽区宽度 ··· 41
3.6　一维正向电流/电压特性 ·· 43
　　3.6.1　小注入条件 ··· 43
　　3.6.2　大注入条件 ··· 45
　　3.6.3　测量电流/电压特性的示例 ··· 46
3.7　多维正向电流/电压特性 ·· 47
　　3.7.1　表面复合对 p^+n 二极管外部电流的影响 ······································· 47
　　3.7.2　电场强度对非自对准台面型 p^+n 二极管的影响 ···························· 49
3.8　结击穿 ··· 52
3.9　总结 ·· 52
参考文献 ··· 52

第4章　光子回收效应 ·· **55**

4.1　引言 ·· 55
4.2　光子回收现象的分类 ·· 56
4.3　本征光子回收 ·· 58
4.4　本征光子回收对正偏 GaN p-n 结二极管的影响 ······································ 60
4.5　自热效应对正偏 GaN p-n 结二极管的影响 ··· 62
4.6　非本征光子回收对正偏 GaN p-n 结二极管的影响 ··································· 63
4.7　非本征光子回收的可能模型 ··· 67
4.8　总结 ·· 68

参考文献 ·· 68

第 5 章　体块单晶生长 ·· **71**
5.1　引言 ·· 71
5.2　HVPE 法生长 GaN ·· 72
　　5.2.1　HVPE 法生长 GaN 的机制 ·· 72
　　5.2.2　GaN HVPE 法生长中的掺杂 ·· 73
　　5.2.3　GaN 的横向外延生长 ·· 73
5.3　高压氮溶液生长 GaN ·· 74
5.4　钠助溶剂生长 GaN ·· 74
5.5　氨热法生长 GaN ·· 74
5.6　升华法生长 SiC 单晶 ·· 75
　　5.6.1　SiC 的升华法生长原理 ·· 75
　　5.6.2　升华法生长 SiC 单晶中的掺杂 ·· 76
5.7　高温化学气相沉积法生长 SiC 单晶 ·· 76
5.8　溶液生长法生长 SiC 单晶 ·· 77
5.9　总结 ·· 77
参考文献 ·· 78

第 6 章　外延生长 ·· **82**
6.1　引言 ·· 82
6.2　GaN 金属有机化学气相沉积 ·· 82
6.3　二维成核理论 ·· 85
6.4　BCF 理论 ·· 86
6.5　4H-SiC 的化学气相沉积 ·· 87
6.6　4H-SiC 的化学气相沉积沟槽填充 ·· 92
6.7　总结 ·· 94
参考文献 ·· 95

第 7 章　制作工艺 ·· **99**
7.1　引言 ·· 99
7.2　刻蚀 ·· 99

  7.2.1 ICP 刻蚀 ·········· 100
  7.2.2 湿法化学刻蚀 ·········· 100
 7.3 离子注入 ·········· 102
  7.3.1 离子注入 GaN ·········· 102
  7.3.2 铝离子注入 4H-SiC ·········· 103
  7.3.3 氮离子和磷离子注入 4H-SiC ·········· 106
 7.4 扩散 ·········· 106
  7.4.1 SiC 中硼扩散的历史背景 ·········· 107
  7.4.2 双子晶格扩散建模 ·········· 108
  7.4.3 半原子模拟 ·········· 110
 7.5 氧化 ·········· 112
  7.5.1 GaN 的热氧化 ·········· 112
  7.5.2 4H-SiC 的热氧化 ·········· 112
 7.6 金属化 ·········· 113
  7.6.1 与 GaN 的欧姆接触 ·········· 113
  7.6.2 与 4H-SiC 的欧姆接触 ·········· 113
 7.7 钝化 ·········· 114
 7.8 总结 ·········· 114
 参考文献 ·········· 114

第 8 章 金属半导体接触和单极功率二极管 ·········· 123
 8.1 引言 ·········· 123
 8.2 肖特基势垒的降低 ·········· 125
 8.3 正向偏置的肖特基结 ·········· 125
 8.4 基于扩散理论的正向电流/电压特性 ·········· 127
 8.5 基于 TED 理论的正向电流/电压特性 ·········· 129
 8.6 基于 TFE 理论的反向电流/电压特性 ·········· 129
 8.7 纯 SBD ·········· 132
  8.7.1 纯 GaN SBD ·········· 132
  8.7.2 纯 4H-SiC SBD ·········· 134
 8.8 缓变 AlGaN SBD ·········· 135
 8.9 带有 p$^+$型薄层的 4H-SiC SBD ·········· 135

8.10 屏蔽平面 SBD ·········· 138
 8.10.1 GaN 混合型 p-n 肖特基二极管 ·········· 138
 8.10.2 4H-SiC JBS 二极管 ·········· 139
8.11 总结 ·········· 141
参考文献 ·········· 141

第9章 金属绝缘体半导体电容器和单极功率开关器件 ·········· **145**

9.1 引言 ·········· 145
9.2 MIS 电容器 ·········· 146
 9.2.1 理想的 MIS 电容器 ·········· 146
 9.2.2 绝缘介质和固定电荷对 MIS 电容器的影响 ·········· 148
9.3 AlGaN/GaN 异质结构 ·········· 149
9.4 GaN、AlN、4H-SiC 和代表性绝缘介质的能带阵容 ·········· 150
9.5 GaN HFET ·········· 151
 9.5.1 GaN MIS HFET ·········· 151
 9.5.2 GaN MESFET ·········· 155
 9.5.3 GaN p^+ 栅极 HFET ·········· 156
9.6 4H-SiC JFET ·········· 158
9.7 MISFET ·········· 159
 9.7.1 平面 MISFET ·········· 161
 9.7.2 沟槽 MISFET ·········· 165
 9.7.3 SJ MISFET ·········· 168
9.8 总结 ·········· 169
参考文献 ·········· 169

第10章 双极功率二极管和功率开关器件 ·········· **176**

10.1 引言 ·········· 176
10.2 一维 p-n 结二极管的优化设计 ·········· 179
10.3 具有非均匀掺杂漂移层的 GaN p-n 结二极管 ·········· 182
10.4 4H-SiC p-i-n 二极管 ·········· 184
 10.4.1 已报道的 4H-SiC p-i-n 二极管结果 ·········· 184
 10.4.2 正偏 4H-SiC p-i-n 二极管的存储电荷 ·········· 185

10.4.3 4H-SiC p-i-n 二极管的反向恢复 ············· 186
10.5 n-p-n 双极晶体管 ························· 186
　　10.5.1 集电极层设计 ······················· 187
　　10.5.2 基极层设计 ························· 189
　　10.5.3 二次击穿的临界集电极电流密度 ········· 189
　　10.5.4 GaN BJT ······························ 189
　　10.5.5 SiC BJT ······························ 190
10.6 肖克利二极管 ··························· 190
　　10.6.1 肖克利二极管的反向阻断 ··············· 191
　　10.6.2 肖克利二极管的正向阻断 ··············· 191
10.7 SiC 晶闸管 ····························· 193
10.8 SiC 绝缘栅型晶闸管 ····················· 193
10.9 总结 ··································· 195
参考文献 ····································· 195

第 11 章 边缘终端 ······························· **199**
11.1 引言 ··································· 199
11.2 GaN 功率器件的 MFP ····················· 203
　　11.2.1 不带保护环的 MFP ··················· 203
　　11.2.2 保护环辅助 MFP ····················· 204
11.3 用于 4H-SiC 功率器件的 SM-JTE ·········· 204
11.4 用于 4H-SiC 功率器件的 CD-JTE ·········· 205
11.5 用于 4H-SiC 功率器件的混合型 JTE ······· 205
11.6 总结 ··································· 206
参考文献 ····································· 207

第 12 章 垂直型 GaN 和 SiC 功率器件可靠性 ········ **209**
12.1 引言 ··································· 209
12.2 HTRB 应力耐受性 ······················· 209
12.3 HTGB 应力耐受性 ······················· 211
12.4 H3TRB 应力耐受性 ······················ 212
12.5 TC 应力耐受性 ························· 213

12.6 HTO 应力耐受性 ………………………………………………… 213
12.7 地面宇宙辐射耐受性 …………………………………………… 213
12.8 总结 ……………………………………………………………… 215
参考文献 ……………………………………………………………… 215

第 1 章

垂直型与横向功率半导体器件

1.1 引言

Si 基电子器件的成功应用促进了功率相对较低（小于 10W）的微电子和纳米电子技术的快速发展（见图 1.1）。即使在处理更高功率的电力电子系统中，Si 基电子器件在近几十年中也发挥着重要作用。然而，Si 基电子器件正在逼近材料基本特性的极限。由于宽禁带半导体具有比 Si 更优越的材料特性，因此作为新一代功率器件用半导体材料引发了广泛关注（见 2.7 节）。

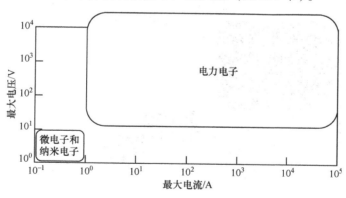

图 1.1 微电子、纳米电子和电力电子器件的面积与最大电压和电流的关系

最具有优势的宽禁带半导体材料是碳化硅（SiC），例如将其应用到轨道交通牵引逆变器中可以显著地降低功率变换损耗[1]。另一种新的宽禁带半导体材料是氮化镓（GaN）。对于开关电源和消费类电子产品，采用 Si 衬底 GaN 外延材料的横向功率器件具有较大的成本效益[2]（见 1.3.1.1 节）。此外，垂直型 GaN 功率器件由于具

有较大的电流处理能力[3]，也吸引了广泛关注。如图 1.2a 和 b 上的箭头所示，SiC 和 GaN 都有望提高器件的额定功率（通过增加芯片中的电流密度）和开关频率，如 2.7 节所述。本章的 1.3 节对垂直型和横向 GaN 和 SiC 功率器件进行了比较。

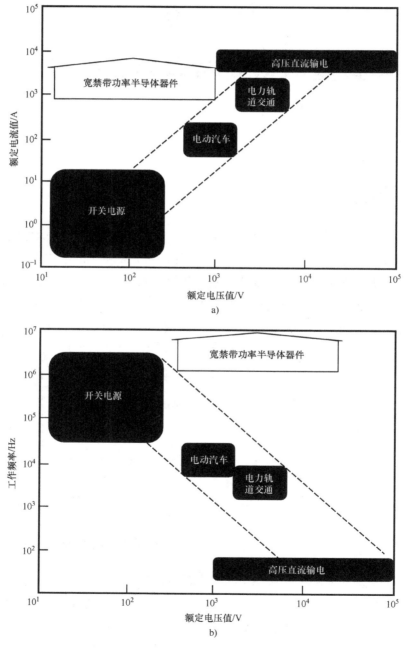

图 1.2 功率半导体器件的应用：a) 额定电流，b) 工作频率

1.2 典型功率半导体器件特性

功率半导体器件在电力电子系统中被用作功率开关和二极管[4]，电能由交流变换到直流（整流器）、直流到直流的转换、直流到交流（逆变器）的转换。二极管是一种两端器件，只允许电流沿一个方向通过（从阳极到阴极），同时阻挡反向电流（直到达到击穿电压 [BV]，见图 1.3）。图 1.4 所示为一个典型的三相电源逆变器的电路图，使用了 6 个开关管和 6 个二极管，输出为交流电压和三相平衡负载[5]。当负载是电感时（例如：电动机），二极管被用作反向导电器件（因为大多数半导体开关只在一个方向传导电流）。

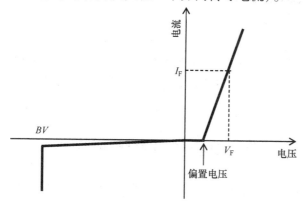

图 1.3 功率半导体二极管的电流/电压特性示意图
（I_F：正向电流；V_F：正向压降；BV：击穿电压）

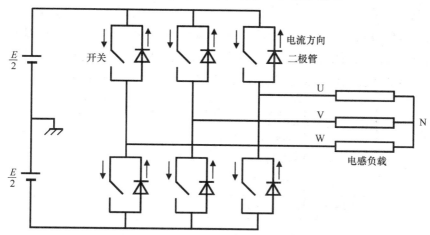

图 1.4 三相电源逆变器的电路图

功率半导体二极管的正向特性表现出正向导通特性（见图 1.3），功率半导体开关器件表现出零偏置电压导通特性（见图 1.5a）或非零偏置电压导通特性（见图 1.5b）；在这两种情况下，当正向电流为 I_F 时，正向电压为 V_F。

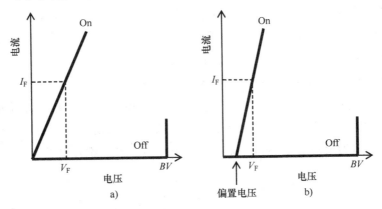

图 1.5 a）功率半导体开关器件的零偏置电压和 b）非零偏置电压导通特性示意图（I_F：正向电流；V_F：正向电压；BV：击穿电压）

1.3 垂直型与横向单极功率半导体器件

半导体中有两种载流子参与导电，即带负电荷的电子和带正电荷的空穴。因此，功率半导体器件可以分为单极器件和双极器件。对于单极器件，通过电子或空穴导电；在双极器件中，由电子和空穴两种载流子同时参与导电。如图 1.6 所示为横向功率半导体器件（单极和双极器件）分类表，相关文献已经报道了 Si 基 GaN、SiC 基 GaN、蓝宝石基 GaN、GaN 基 GaN 及 SiC 基 SiC 横向单极功率开关器件和横向单极功率二极管，而横向双极功率开关器件应用较少。因此，本节将比较单极功率开关器件和功率二极管的垂直和横向器件。

	横向功率开关器件	横向功率二极管
单极器件	• Si基GaN • SiC基GaN • GaN基GaN • 蓝宝石基GaN • SiC基SiC	
双极器件	不常用	不常用

图 1.6 横向功率半导体器件（单极和双极器件）分类表

1.3.1 垂直型和横向单极功率开关器件

图 1.7 显示了面积为 1cm² 的横向和垂直型单极功率开关器件。单极功率开关器件表现出零偏置电压特性，如图 1.5a 所示，其中特征导通电阻 $R_{on}A$ 被定义为 $V_F A/I_F$。因为 $R_{on}A$ 不包括特征衬底电阻 $R_{sub}A$（见图 1.7a），所以横向单极功率开关器件比垂直型单极功率开关器件更具有优势。

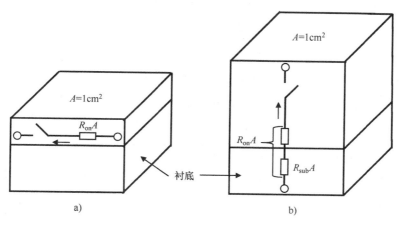

图 1.7 面积为 1cm² 的 a) 横向和 b) 垂直型单极功率开关器件示意图，其中 $R_{on}A$ 和 $R_{sub}A$ 分别表示特征导通电阻和特征衬底电阻

1.3.1.1 垂直型与横向 GaN 单极功率开关器件

图 1.8 所示为研制的横向 GaN 单极功率开关器件室温特征导通电阻（$R_{on}A$）与额定电压的函数关系[2]。如图 1.8 所示，GaN 和 SiC 衬底的 $R_{sub}A$ 值计算如下：GaN 衬底的电阻率和厚度分别为 0.018Ω·cm 和 600μm[6]，而 SiC 衬底的电阻率和厚度分别为 0.015~0.02Ω·cm 和 350~500μm[7]，从而计算得到 GaN 和 SiC 衬底的 $R_{sub}A$ 值分别为 1.1mΩ·cm²（见图 1.8 虚线）和 0.53~1.4mΩ·cm²（见图 1.8 点线）。此外，只要额定电压小于 200V，横向 GaN 单极功率开关器件的 $R_{on}A$ 值随额定电压的增加而增加，但不超过 GaN 和 SiC 衬底的 $R_{sub}A$ 值。如 1.1 节所述，之前制备的 Si 衬底横向 GaN 单极开关器件比 GaN 衬底型器件更便宜、面积更大[2]，横向 GaN 单极功率开关器件更适合于低成本和低电压应用，如图 1.2 所示的开关电源。

对于超过 400V 额定电压的横向 Si 衬底 GaN 单极功率开关器件的最大电流为 70A（面积 A 为 0.29cm²）[8]（见图 1.9），自 2009 年以来没有更大电流的报

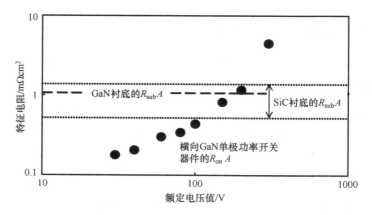

图 1.8 研制的横向 GaN 单极功率开关器件室温特征导通电阻（$R_{on}A$）与额定电压的函数关系[2]，GaN 衬底（虚线）和 SiC 衬底（点线）的特征衬底电阻（$R_{sub}A$）与电压额定值的关系[6,7]

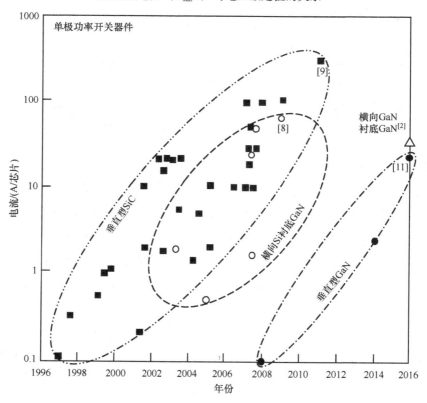

图 1.9 横向 Si 衬底 GaN（空心圆），横向 GaN 衬底 GaN 器件（空心三角形），垂直型 GaN（实心圆）和垂直型 SiC（实心正方形）单极功率开关器件的电流和报道年份示意图

第1章 垂直型与横向功率半导体器件

道,这个值似乎是一个极限报道值。较大面积的横向 Si 衬底 GaN 单极功率开关器件,由于受到异质衬底上 GaN 层中高密度的晶格缺陷(见 2.2.3 节)和残余应力的影响而使其发展受到限制。

另一方面,垂直型 GaN 单极功率开关器件与垂直型 SiC 单极功率开关器件具有相同的趋势($1.0cm^2$ 面积电流达到 350A[9]),但其发展滞后 10 年左右。通常,垂直型功率开关器件具有高耐压的优点,因此可以避免使用横向功率开关器件中的横跨金属化层[10]。垂直型 GaN 单极功率开关器件[11]报道的电流达到 23A(面积 A 为 $0.023cm^2$),预估可以达到 350A。很明显,垂直型 GaN 单极功率开关器件适用于大电流和高电压环境下工作。需要注意的是,采用新开发的横向 GaN 衬底单极功率开关器件面积为 $0.040cm^2$ 时获得了 32A 的电流[12],因此 GaN 单极功率开关器件的最大电流将来可能会变得更大。然而,采用 GaN 衬底的横向器件相较于 Si 衬底 GaN 横向器件在成本上没有优势。

1.3.1.2 垂直型和横向 SiC 单极功率开关器件

图 1.10 显示了横向[13-19]和垂直型[20-22] SiC 单极功率开关器件的室温特征导通电阻 $R_{on}A$ 与击穿电压的函数关系。报道的横向 SiC 单极功率开关器件 $R_{on}A$ 比

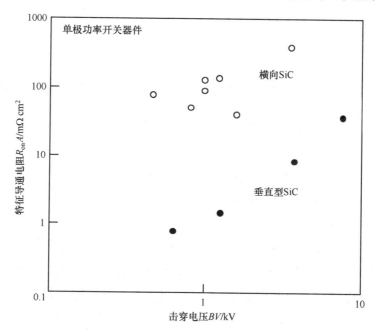

图 1.10 报道的横向(空心圆)[13-19]和垂直型(实心圆)[20-22] SiC 单极功率开关器件的室温特征导通电阻 $R_{on}A$ 与击穿电压的函数关系

垂直型 SiC 器件高 10 倍以上。虽然在 2005 年提出了改善 $R_{on}A/BV$ 关系的方法[23]，但其效果还有待证明。因此，垂直型 SiC 单极功率开关器件适合于大电流和高电压环境下工作，这与 GaN 单极功率开关器件的性质相似（见 1.3.1.1 节）。

1.3.2 垂直型和横向单极功率二极管

由于横向 SiC 单极功率二极管的论文较少，因此横向结构只考虑 GaN 单极功率二极管。与功率开关器件不同，功率二极管没有控制端。因此，对于横向单极功率二极管，可与多个二极管并联（见图 1.11）；例如文献 [25] 中多达 8 个并联二极管（由 AlGaN/GaN 异质结[24]构成）。根据文献 [26] 报道，横向单极功率二极管中 3 个并联的 Si 衬底 AlGaN/GaN 二极管击穿电压和芯片尺寸分别为 600V 和 3.4mm×4mm，最大芯片电流为 20A。

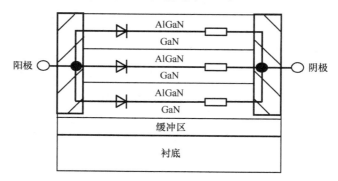

图 1.11　3 个并联 AlGaN／GaN 异质结二极管组成的横向单极功率二极管示意图

面积为 1cm² 的横向和垂直型单极功率二极管如图 1.12 所示。单极功率二极管具有如图 1.3 所示的导通电压特性，其中微分特征导通电阻 $R_{on}A^{diff}$ 定义为 dV_FA/dI_F。横向单极功率二极管比垂直型单极功率二极管更有优势，这里 $R_{on}A^{diff}$ 不包括 $R_{sub}A$。然而，与横向单极功率开关器件不同，额定电压小于 200V 的横向单极功率二极管尚未见报道。

上述三个并联额定电压为 600V 的 Si 衬底 AlGaN/GaN 二极管[26]与 790V、50A 的垂直型 GaN 单极功率二极管（阳极面积为 3mm×3mm）[27]电流密度和电压特性关系如图 1.13 所示。由于垂直型 GaN 单极功率二极管的芯片尺寸在文献 [27] 中没有说明，所以计算其电流密度时，假设了两个可能的尺寸（最小：3.4mm×3.4mm；最大：4mm×4mm）。即使假设较大的芯片面积，垂直型 GaN 单极功率二极管的电流密度也大于三个并联的 AlGaN/GaN 二极管（见图 1.13）。

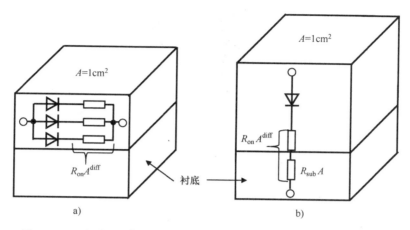

图 1.12 面积为 1cm² 的 a) 横向和 b) 垂直型单极功率二极管示意图，其中 $R_{on}A^{diff}$ 和 $R_{sub}A$ 分别表示微分特征导通电阻和特征衬底电阻

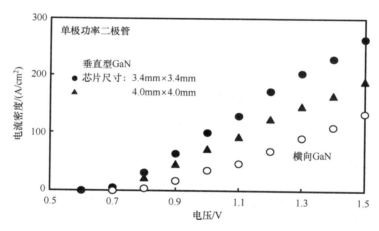

图 1.13 大电流（> 20A）横向 GaN（空心圆：芯片尺寸为 3.4mm×4mm）[26] 和垂直型 GaN（实心圆：最小芯片尺寸为 3.4mm×3.4mm；实心三角形：最大芯片尺寸为 4mm×4mm）的电流密度和电压特性关系

自 2011 年以来，横向 GaN 单极功率二极管的最大芯片电流[26]并没有增加，但垂直型 GaN 单极功率二极管的最大芯片电流仍在增加[27]，并且按照垂直型 SiC 单极功率二极管的发展趋势在发展，但其发展延后约 10 年[28]（见图 1.14）。

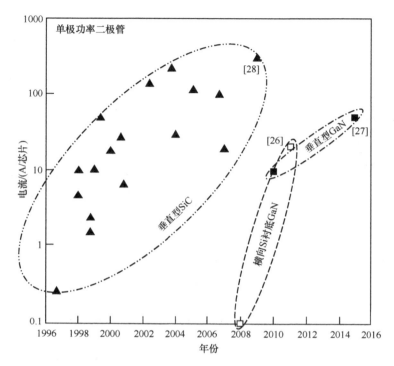

图 1.14　已报道的横向 Si 衬底 GaN（空心正方形）、垂直型 GaN（实心正方形）和垂直型 SiC（实心三角形）单极功率二极管电流

1.4　总结

本章以三相电源逆变器为例，简要说明了功率半导体开关器件和二极管的功能。作为下一代功率开关器件用半导体材料，SiC 和 GaN 是最有前途的宽禁带半导体材料。通过垂直型和横向器件比较了两种材料制备的单极功率开关器件和二极管，当额定电压小于 200V 时，横向 GaN 单极功率开关器件具有较低的 $R_{on}A$ 优势；当额定电压大于 400V 时，垂直型 GaN 和 SiC 功率开关器件和功率二极管比横向 GaN 器件能获得更大的电流和更高的电流密度，根据这些结论，以下的章节只讨论垂直型功率半导体器件。

参 考 文 献

[1] Hamada, K., et al., "3.3 kV/1,500A Power Modules for the World's First All-SiC Traction Inverter," *Japanese Journal of Applied Physics*, Vol. 54, 2015, pp. 04DP07-1–04DP07-4.

[2] http://epc-co.com/epc/Products/eGaNFETsandICs.aspx.

[3] https://arpa-e.energy.gov/?q=slick-sheet-project/vertical-gan-transistors.

[4] Shockley, W., "The Theory of p-n Junctions in Semiconductors and p-n Junction Transistors," *Bell System Technical Journal*, Vol. 28, No. 3, 1949, pp. 435–489.

[5] Blaabjerg, F., and J. K. Pedersen, "Optimized Design of a Complete Three-Phase PWM-VS inverter," *IEEE Transactions on Power Electronics*, Vol. 12, No. 3, 1997, pp. 567–577.

[6] Yoshida, T., et al., "Preparation of 3-inch Freestanding GaN Substrates by Hydride Vapor Phase Epitaxy with Void-Assisted Separation," *Phyica Status Solidi A*, Vol. 205, No. 5, 2008, pp. 1053–1055.

[7] https://www.wolfspeed.com/index.php/downloads/dl/file/id/888/product/0/materials_catalog.pdf.

[8] Kambayashi, H., et al., "Enhancement-Mode GaN Hybrid MOS-HFETs on Si Substrates with over 70-A Operation," *International Symposium on Power Semiconductor Devices and ICs*, Barcelona, June 14–18, 2009, pp. 21–24.

[9] Nakamura, T., et al., "High-Performance SiC Power Devices and Modules with High Temperature Operation," *International Symposium on VLSI, Automation and Test*, Hsinchu, April 25–28, 2011, pp. 1–2.

[10] Ueda, D., "Renovation of Power Devices by GaN-Based Materials," *International Electron Devices Meeting*, Washington, D.C., June 7–9, 2015, pp. 422–425.

[11] Oka, T., et al., "Over 10-A Operation with Switching Characteristics of 1.2 kV-Class Vertical GaN Trench MOSFETs on a Bulk GaN Substrate," *International Symposium on Power Semiconductor Devices and ICs*, Prague, June 12–16, 2016, pp. 459–462.

[12] Handa, H., et al., "High-Speed Switching and Current-Collapse-Free Operation by GaN Gate Injection Transistors with Thick GaN buffer on Bulk GaN Substrates," *International Electron Devices Meeting*, San Francisco, Dec. 5–7, 2016, pp. 256–259.

[13] Okamoto, M., et al., "Lateral RESURF MOSFET Fabricated on 4H-SiC ($000\bar{1}$) C-Face," *IEEE Electron Device Letters*, Vol. 25, No. 6, 2004, pp. 405–407.

[14] Fujikawa, K., et al., "800 V 4H-SiC RESURF-Type Lateral JFETs," *IEEE Electron Device Letters*, Vol. 25, No. 12, 2004, pp. 790–791.

[15] Banerjee, S., et al., "Robust, 1000-V, 130-mΩcm^2, Lateral Two-Zone RESURF MOSFETs in 6H-SiC JFETs," *International Symposium on Power Semiconductor Devices and ICs*, Santa Fe, June 4–7, 2002, pp. 69–72.

[16] Zhang, Y., et al., "1,000-V 9.1-mΩcm^2 Normally off 4H-SiC Lateral RESURF JFET for Power Integrated Circuit Applications," *IEEE Electron Device Letters*, Vol. 28, No. 5, 2007, pp. 404–407.

[17] Kimoto, T., et al., "Design and Fabrication of RESURF MOSFETs on 4H-SiC(0001), (11$\bar{2}$0), and 6H-SiC(0001)," *IEEE Transactions on Electron Devices*, Vol. 52, No. 1, 2005, pp. 112–117.

[18] Noborio, M., J. Suda, and T. Kimoto, "4H-SiC Lateral Double RESURF MOSFETs with Low on Resistance," *IEEE Transactions on Electron Devices*, Vol. 54, No. 5, 2007, pp. 1216–1223.

[19] Lee, W.-S., et al., "Design and Fabrication of 4H-SiC Lateral High-Voltage Devices on a Semi-Insulating Substrate," *IEEE Transactions on Electron Devices*, Vol. 59, No. 3, 2012, pp. 754–760.

[20] Nakamura, T., et al., "High Performance SiC trench Devices with Ultra-Low R_{on}," *International Electron Devices Meeting*, Washington, DC, June 5–7, 2011, pp. 599–601.

[21] Harada, S., et al., "3.3-kV-Class 4H-SiC MeV-Implanted UMOSFET with Reduced Gate Oxide Field," *IEEE Electron Device Letters*, Vol. 37, No. 3, 2016, pp. 314–316.

[22] Kawahara, K., et al., "6.5-kV Schottky-Barrier-Diode-Embedded SiC-MOSFET for Compact Full-Unipolar Module," *International Symposium on Power Semiconductor Devices and ICs*, Sapporo, May 28–June 1, 2017, pp. 41–44.

[23] Baliga, B. J., *Silicon Carbide Power Devices*, Singapore: World Scientific, 2005, pp. 471–472.

[24] Ishida, H., et al., "Unlimited High Breakdown Voltage by Natural Super Junction of Polarized Semiconductor," *IEEE Electron Device Letters*, Vol. 29, No. 10, 2008, pp. 1087–1089.

[25] Terano, A., et al., "GaN-Based Multi-Two-Dimensional-Electron-Gas-Channel Diodes on Sapphire Substrates with Breakdown Voltage of over 3 kV," *Japanese Journal of Applied Physics*, Vol. 54, 2015, pp. 06650-1–06650-5.

[26] Shibata, D., et al., "GaN-Based Multijunction Diode with Low Reverse Leakage Current Using p-Type Barrier Controlling Layer," *International Electron Devices Meeting*, San Francisco, Dec. 5–7, 2011, pp. 587–590.

[27] Tanaka, N., et al., "50-A Vertical GaN Schottky Barrier Diode on a Freestanding GaN Substrate with Blocking Voltage of 790V," *Applied Physics Express*, Vol. 8, 2015, pp. 071001-1–071001-3.

[28] Nakamura, T., et al., "Development of SiC Diodes, Power MOSFETs and Intelligent Power Modules," *Physica Status Solidi A*, Vol. 206, No. 10, 2009, pp. 2403–2416.

第 2 章

GaN 和 SiC 的物理性质

2.1 引言

如 1.1 节所述，宽禁带半导体如 GaN 和 SiC 相较于 Si，因其优越的材料特性而备受关注，然而，GaN 和 SiC 在自然界中并不是大量存在的。1969 年，Maruska 和 Tietjen 首次以粉末形式合成了 GaN 材料[1]，而 SiC 的首次合成（1824 年）据说是偶然的：Berzelius 在尝试制造金刚石时得到了 SiC 材料[2]。

GaN 或 SiC 功率器件制造中通过衬底生长 GaN 或 SiC 层（在 1.3 节中介绍）。由于 GaN 衬底难以制备，迫使早期研究人员使用蓝宝石作为衬底，因为蓝宝石的热膨胀系数与 GaN 相对接近。从 1986 年开始，Akasaki、Amano 和他们的同事通过在 GaN 层和蓝宝石衬底之间插入 AlN 成核层提高 GaN 外延层的质量[3,4]。1988 年，他们实现了 GaN 的 p 型掺杂，在 GaN 中引入了空穴[5]（带正电荷的载流子［见 1.3 节］）。第一次开发了 GaN 发光器件之后（Nakamura 等人在 1991 年[6]制造的 p-n 结蓝光发光二极管），1993 年[7]首次报道了蓝宝石衬底上的 GaN 电子器件。然而，在文献［8］报道的生长温度条件下，蓝宝石衬底上生长的 GaN 外延层存在高密度晶体缺陷，大约有 13% 的晶格失配（2.2.3 节中所述的位错）。1999 年，通过在蓝宝石衬底上生长 250~300μm 厚的 GaN 膜，能够形成 2in⊖ 的 GaN 衬底[9]。从那时开始，国际上就已经可以制造出垂直型 GaN 功率器件[10]，并且研究报道了具有低密度缺陷的 GaN 外延层电学特性[11]。

与 GaN 相比，自 1978 年 Tairov 和 Tsvetkov 开发出一种称为改良 Lely 的籽晶

⊖ 1in=0.0254m。——编辑注

升华法以后，主要用作生长 SiC 衬底[12]。需要升华是因为碳（C）在 Si 溶液中的溶解度很低（例如，在 2073~2273K 时，只有 0.1% 的碳可以溶解在硅中[13]）。与 GaN 中的 p 型掺杂相比，SiC 中的 p 型掺杂并不困难。但是，由于 SiC 结晶有 250 多种变体（所谓的多型体）[14]，因此很难生长 SiC 层。1987 年，Kuroda 等人提出了一种称为阶梯控制外延的技术[15]，该技术可以在 SiC 层中复制衬底上已生长的相同多型体。请注意，外延一词来自两个希腊词：Epi 意思是在上方，taxis 意思是排列，是指相对于衬底晶体结构具有明确方向的晶体覆盖层（见第 6 章）。由于上述两种技术，制造和应用 SiC 功率器件在 20 世纪 90 年代初变得成熟[16]。

鉴于上述关于 GaN 和 SiC 的历史，本章首先将比较 GaN 和 SiC 的晶体结构，包括晶体缺陷，然后描述它们的能带、杂质掺杂、载流子迁移率和碰撞电离。由于垂直型 GaN 功率开关器件使用了 AlGaN/GaN 异质结构[17-19]，因此本章也将描述 AlN 的晶体结构和能带。2.7 节将介绍功率器件常用的性能指标。

2.2 晶体结构

AlN、GaN 和 SiC 是化合物半导体，其化学计量可为 50% 的 Al、Ga 或 Si 和 50% 的 N 或 C。注意，硅锗（$Si_{1-x}Ge_x$）不是一种化合物半导体，而是一种合金半导体，因为 x 可以取 0~1 之间的任何值。Al、Ga、N、Si、C 原子基态的电子结构如下：

$$_{13}Al: 1s^2\ 2s^2\ 2p^6\ 3s^2\ 3p^1 \tag{2.1a}$$

$$_{31}Ga: 1s^2 2s^2\ 2p^6\ 3s^2\ 3p^6\ 3d^{10}\ 4s^2 4p^1 \tag{2.1b}$$

$$_{7}N: 1s^2\ 2s^2\ 2p^3 \tag{2.1c}$$

$$_{14}Si: 1s^2\ 2s^2\ 2p^6\ 3s^2\ 3p^2 \tag{2.1d}$$

$$_{6}C: 1s^2\ 2s^2 2p^2 \tag{2.1e}$$

其中下标和上标分别表示原子序数和轨道电子数。Al 和 Ga 的最外层都有 3 个价电子，而 N 的最外层有 5 个价电子。AlN 和 GaN 因此被称为Ⅲ-Ⅴ化合物半导体。SiC 被称为Ⅳ-Ⅳ化合物半导体，因为 Si 和 C 的最外层都有 4 个价电子。

2.2.1 AlN 和 GaN 的晶体结构

AlN 和 GaN 可以形成以下任何一种晶体结构：岩盐（见图 2.1）、闪锌矿

（见图2.2a）或纤锌矿（见图2.2b）。仅在非常高的压力下可以观察到岩盐结构：23GPa的AlN[20]，52GPa的GaN[21]。在自然条件下，AlN和GaN的热力学稳定结构均为纤锌矿。各种化合物半导体的能量差表示在图2.3中（使用波函数和经典轨道半径[22]计算获得）。与诸如GaP和GaAs之类的闪锌矿稳定Ⅲ-Ⅴ化合物半导体相反，AlN和GaN都是纤锌矿稳定Ⅲ-Ⅴ化合物半导体。注意，从图2.3中，尚不清楚SiC是闪锌矿稳定还是纤锌矿稳定的Ⅳ-Ⅳ化合物半导体。根据2.2.2节所述，闪锌矿SiC通常出现在相对较低的温度下。

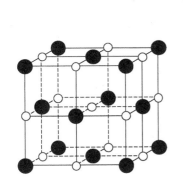

图2.1 岩盐晶体结构
（实心圆：铝和/或镓原子；
空心圆：氮原子）

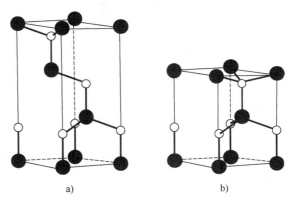

图2.2 a）闪锌矿和b）纤锌矿晶体结构
（实心圆：铝和/或镓原子；空心圆：氮原子）

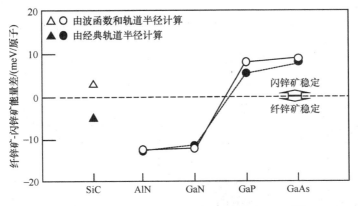

图2.3 计算得到的纤锌矿-闪锌矿结构能量差[22]

纤锌矿结构（见图2.2b）的平面图和透视图如图2.4所示。在这样的六方棱柱晶体中，可以用米勒-布拉维指数（hkil）[23]表示晶面。

$$h + k + i = 0 \tag{2.2}$$

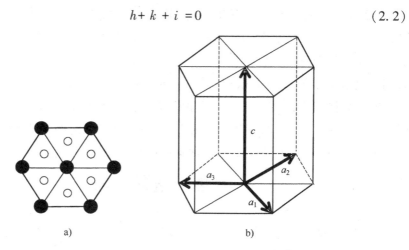

图 2.4 纤锌矿结构的平面 a) 和透视图 b)
（实心圆：铝和/或镓原子；空心圆：氮原子）

米勒-布拉维指数通过以下步骤确定：
- 确定平面的 a_1、a_2、a_3 和 c 轴的截距。
- 形成截距的倒数。
- 找出与截距比例相同的最小整数集。
- 如果是负的，就用横线表示。

典型纤锌矿结构中的晶面如图 2.5 所示。图 2.5a 所示的（0001）平面是 GaN 功率器件[9]最常用的衬底表面，具有 Al 或 Ga 极性（见图 2.6a）。（000$\bar{1}$）具有 N 极性 GaN[24]（见图 2.6b），（10$\bar{1}$0）是所谓的 m 平面 GaN[25]，（11$\bar{2}$0）是 a 平面 GaN[26]，（0001）AlN[27]，（10$\bar{1}$0）AlN[28] 和 （0001）AlGaN[29]。

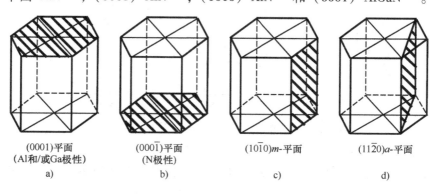

图 2.5 典型纤锌矿结构（见图 2.2b 和图 2.4）中的晶面

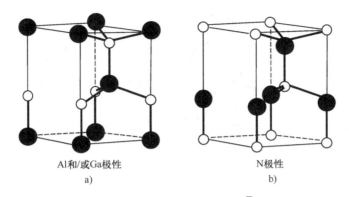

图 2.6 纤锌矿结构中 a)（0001）和 b)（000$\bar{1}$）平面的原子排列
（实心圆：铝和/或镓原子；空心圆：氮原子）

2.2.2 SiC 的晶体结构

晶体类型用晶胞和晶体系统中的 Si-C 堆叠层数表示（例如 C 代表立方，H 代表六角形）。常见的晶体类型（3C-SiC、4H-SiC、6H-SiC）和 2H-SiC 的原子排列如图 2.7 所示，其中 A，B 和 C 为六角形密堆积结构中的占据位置（见图 2.8）。3C-SiC（闪锌矿 SiC）是一种低温稳定的晶体，在 2300K 以上会转变为六角形[30]。2H-SiC（纤锌矿 SiC）在高温下也不稳定。6H-SiC 中的电子转移受晶向影响较大（见 2.5 节），所以在功率半导体中 4H-SiC 优于 6H-SiC。

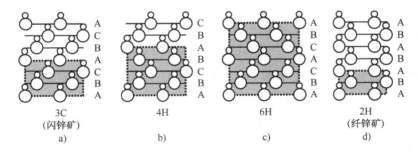

图 2.7 常见晶体的截面示意图 a) 3C-SiC，b) 4H-SiC，c) 6H-SiC 和
d) 2H-SiC，大空心圆和小空心圆分别表示硅原子和碳原子

根据上述定义，闪锌矿和纤锌矿的 GaN（或 AlN）可以表示为 3C-GaN 和 2H-GaN（或 AlN）。为了方便，下文中的 GaN（或 AlN）是指最稳定的纤锌矿结构。

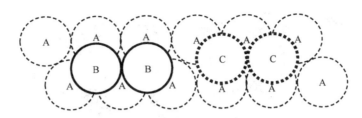

图 2.8 六边形密排结构中占位点原子（A、B 和 C）的平面示意图

2.2.3 晶体缺陷

由于有些晶体缺陷具有电学活性，因此是功率器件关注的重点。晶体缺陷通常可分为四种类型：零维（即点）、一维（即线）、二维（即面）和三维（即体）缺陷。

点缺陷影响载流子寿命（见第 3 章），因此影响双极器件中的电导调制（见 3.7 节）。两种基本类型的自然点缺陷[31]为空位（也称为肖特基缺陷，即晶格位置缺失原子）和间隙（也称为费仑克尔缺陷，即晶格位置之间附加原子）。在热平衡时，空位（C_V）和间隙（C_I）的浓度取平衡值（C_V^*，C_I^*），其分别由自由能 F_V 和 F_I 确定[32,33]为

$$C_V^* = C_{sV} \exp(-F_V/kT) \tag{2.3a}$$

$$C_I^* = C_{sI} \exp(-F_I/kT) \tag{2.3b}$$

式中，C_{sV} 和 C_{sI} 分别为空位和间隙允许存在位置的浓度；k 为玻尔兹曼常数；T 是绝对温度。在硅材料中，$C_{sV} = 2.0 \times 10^{23} cm^{-3}$，$F_V = 2.0 eV$，$C_{sI} = 5.0 \times 10^{22} cm^{-3}$，$F_I = 2.36 eV$，已在商业仿真过程中广泛使用[34]。尽管在 GaN 和 4H-SiC 中还没有准确值，但是已用于描述硼在 4H-SiC 中的扩散（见 7.4 节）。除了一般的空位和间隙，GaN 中还存在载流子陷阱[35]，而 4H-SiC 中的两个主要点缺陷[38]为 $Z_{1/2}$[36] 和 $EH_{6/7}$[37]。

线缺陷称为位错，刃型位错（见图 2.9a）和螺型位错（见图 2.9b）是位错的两种主要类型，而混合位错介于二者之间。当引入额外的（或缺失的）原子时，会形成刃型位错。图 2.9a 中位于基面（0001）中的位错称为基面位错（BPD），垂直于（0001）的位错称为螺旋刃型位错（TED）。BPD 和 TED 之间的区别仅仅在于方向的不同。因此，通常二者可以相互转化。对于螺型位错（见图 2.9b 中的实线），其是螺旋生长的源头（见 6.2、6.4 和 6.5 节）。但是 GaN 和 4H-SiC 的螺旋台阶高度不同。GaN 中是一个 Ga-N 双层，4H-SiC 中是四个 Si-C 双层。由于螺旋位错在 GaN 中充当非辐射复合中心（见 3.4.2 节）[39]，

因此它们直接影响 GaN 发光器件的性能，特别是激光二极管（LD）。使用蓝宝石衬底可将 GaN 螺旋位错的密度降低到 $10^6 cm^{-2}$ 左右，从而提高 GaN 激光二极管的可靠性。GaN 和 4H-SiC 双极功率器件中螺旋位错同样会影响载流子寿命，但它们不会影响单极功率器件中的载流子寿命[38]。

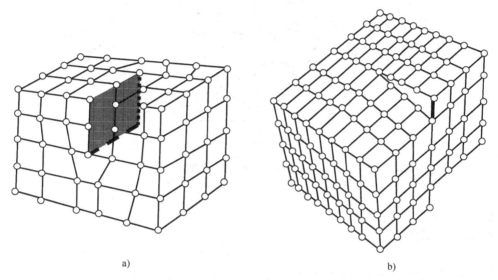

图 2.9 a）刃型位错和 b）螺型位错示意图。在 a）中有一个额外的半平面。
当 a）中的表面为（0001）时，虚线为基面位错，点线为螺旋刃型位错

改变各层周期性顺序的面状缺陷称为堆叠层错（见图 2.10）。GaN 中堆叠层错能量较高（$1.2 J/m^2$）[40]，所以很少出现堆叠层错。此外，研究表明 GaN 中的堆叠层错没有电学活性[41]。但是在 4H-SiC 中，堆叠层错能量较低（$0.014 J/m^2$）[42]，所以常常会出现堆叠层错。

所谓的反转域也称为 GaN 中的面缺陷。反转域与基体生长极性相反。在表面为（0001）的情况下，由于具有 N 极性的 GaN 生长速度比具有 Ga 极性的 GaN 生长速度慢，因此在反转域中会形成 V 形凹坑[43]。反转域与应变相结合会改变压电场[44]，而这些变化可能会影响具有 AlGaN/GaN 异质结构的垂直 GaN 功率开关器件的特性[17-19]。

体缺陷与体积有关，包括空隙、裂纹和/或纳米/微管道。GaN 中的纳米管和 4H-SiC 中的微管是沿晶体生长方向排列的隧道状缺陷（开芯螺型位错），它们可穿透薄膜。在外延生长 4H-SiC 时，微管被解离成多个闭芯螺型位错，导致微管闭合[45,46]。

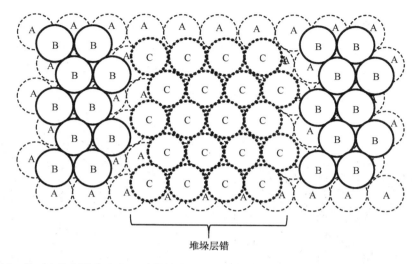

图 2.10　位于占位点原子（A）上方的占位点原子（B）和堆叠层位错点原子（C）的平面示意图

2.3 能带

受原子核库仑势作用的电子只允许存在于某些能级上。当大量原子形成晶体时，每个电子受到的力会改变。泡利不相容原理指出每个允许的电子能级必须是不同的，当有 N 个原子时，初始能级会分裂成 N 个不同的允许能级。由于最初的能级是几电子伏特（eV）的量级，而 N 的量级为 10^{22}，所以 N 个不同能级之间的间隔约为 $10^{-22}\,\mathrm{eV}$。这时能级间隔非常小（与 $10^{-2}\,\mathrm{eV}$ 量级的热能相比），电子很容易在两个能级之间移动。因此，可以将 N 个不同的能级视为连续带（即能带）。

在半导体中，最外层中的价电子完全充满了一个允许的能带（即价带），并且与下一个允许更高能量的带存在能隙（见图 2.11）。当温度高于 0K 时，价带并没有完全填满，因为少数电子拥有足够的热能可以通过激发跨越能隙到下一个允许的能带。由于上面的能带中的电子可以对外加电场做出反应，即产生电流，所以上面的能带被称为导带。导带的下边缘和价带的上边缘通常分别表示为 E_C 和 E_V。当电子被激发到导带中时，在价带中会留下空位。如果施加电场，附近的电子会进入那些空位并产生电流。因此，把价带中的电荷运动看作带正电荷的粒子。这些粒子被称为空穴（在 1.3 节中描述），其概念可以通过类比自来水中的气泡来理解。

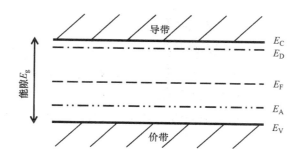

图 2.11 半导体的能带图 [导带 (E_C),施主能级 (E_D), 费米能级 (E_F),受体能级 (E_A) 和价带 (E_V)]

在半导体中,E_C 和 E_V 是晶体动量的函数,并且能隙 E_g 定义为 E_C 和 E_V 之间的差(见图 2.12)。对于直接带隙半导体(例如 AlN 和 GaN),E_C 的最小动量和 E_V 的最大动量是相同的(见图 2.12a)。但是,对于间接带隙半导体(例如硅和 SiC),它们是不同的(见图 2.12 b)。

图 2.12a 中的电子可以在不改变晶体动量的情况下,从价带的最高能量状态转移到导带的最低能量状态。而在图 2.12b 中,如果不改变晶体动量,电子就不能从价带的最高能量状态转移到导带中的最低能量状态。垂直实线箭头表示的能量来自光子(即电磁辐射的量子),而水平虚线箭头表示的能量来自声子(即振动模式的量子力学量子化的一种激发态)。

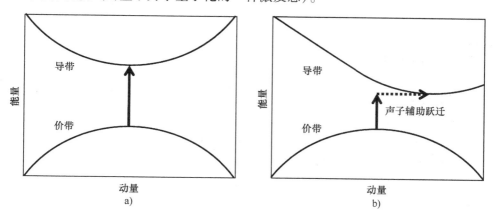

图 2.12 a)直接带隙半导体和 b)间接带隙半导体的能量与晶体动量的对比

E_g 可以通过光学吸收光谱法获得[31],其中光学吸收系数 α(cm^{-1})是最重要的参数。光的强度(I_0)随距离表面的深度(x)指数下降到 $I(x)$,如下所示

$$I(x) = I_0 \exp(-\alpha x) \tag{2.4}$$

在直接带隙半导体中,根据经验,α 与 $(E_{photon} - E_g)^{0.5}$ 成正比,其中 E_{photon} 是光子的能量。图 2.13 所示为文献 [47-49] 中 α 值与 E_{photon} 的函数。在 AlN 和 GaN 中,吸收开始很快,吸收阈值决定了 E_g。而在 4H-SiC 中,即使 E_{photon} 超过 E_g,光吸收也增加缓慢。这种不同的光吸收行为导致光子回收起主导作用或可以忽略(见第 4 章)。

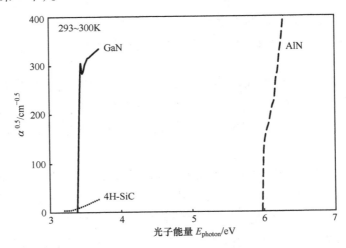

图 2.13　AlN、GaN 和 4H-SiC 中光吸收系数的二次方根与光子能量的函数[47-49]

2.4　杂质掺杂

控制载流子数量最有效的方法是掺入杂质,即向导带提供电子的施主或接受价带中电子(产生空穴)的受主。如果大多数掺杂剂是施主,则半导体称为 n 型;如果大多数掺杂剂是受主,则称半导体为 p 型。

在本征(即非掺杂)半导体中,所有载流子都是由热激发产生的,因此电子浓度 n 和空穴浓度 p 等于本征载流子浓度 n_i。即,质量作用定律:

$$np = n_i^2 \tag{2.5}$$

其中,n_i 为

$$n_i = (N_C N_V)^{0.5} \exp(-E_g/2kT) \tag{2.6a}$$

$$N_C = 2(2\pi m_n kT/h^2)^{1.5} \tag{2.6b}$$

$$N_V = 2(2\pi m_p kT/h^2)^{1.5} \tag{2.6c}$$

式中，N_C 和 N_V 分别为导带和价带的有效态密度；m_n 和 m_p 分别为电子和空穴的有效质量；h 为普朗克常数。

尽管 N_C 和 N_V 都与 $T^{1.5}$ 成正比，但是式（2.6a）中的指数项包含的温度对 n_i 的影响更大。对于 $E_g=3.44\text{eV}$[50] 的 GaN 和 $E_g=3.23\text{eV}$[51] 的 4H-SiC，室温附近每升高 3K，n_i 就会增加一倍。

如文献［52，53］中所示，GaN 和 4H-SiC 的 n_i 与温度的关系分别为

$$n_i^{\text{GaN}} = (4.3\times10^{14}\times T^{1.5}\times 8.9\times10^{15}\times T^{1.5})^{0.5} \times$$
$$\exp(-3.44\times1.6\times10^{-19}/2/1.38\times10^{-23}/T) \quad (2.7\text{a})$$
$$= 2.0\times10^{15}T^{1.5}\exp(-2.0\times10^4/T)(\text{cm}^{-3})$$
$$n_i^{\text{4H-SiC}} = (3.25\times10^{15}\times T^{1.5}\times 4.8\times10^{15}\times T^{1.5})^{0.5} \times$$
$$\exp(-3.23\times1.6\times10^{-19}/2/1.38\times10^{-23}/T) \quad (2.7\text{b})$$
$$= 3.9\times10^{15}T^{1.5}\exp(-1.9\times10^4/T)(\text{cm}^{-3})$$

它们与硅材料中 n_i 随温度的变化如图 2.14 所示。例如，在 150℃ 时，Si 中的 n_i 达到 10^{13}cm^{-3}，这几乎是最低的掺杂水平。而在 GaN 和 4H-SiC 中，即使在 250℃ 下，n_i 也远远低于该水平。较低的 n_i 使得 GaN 和 4H-SiC 功率器件能够在更高的温度下工作。

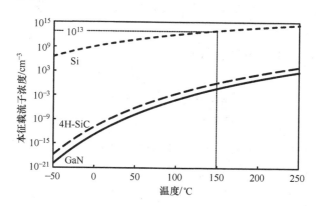

图 2.14　GaN、4H-SiC 和 Si 中本征载流子浓度与温度的关系

2.4.1　n 型掺杂

GaN 中的施主杂质常用硅，而在 4H-SiC 中常用氮或磷。电子从价带激发到导带需要大量的能量（超过 E_g），而将电子从施主激发到导带只需要少量的能量。这要求施主能级 E_D 恰好低于 E_C（见图 2.11）。据报道，硅在 GaN[51] 中的

最小电离能（即 E_C-E_D）为 0.02eV，4H-SiC 中的氮为 0.059eV[54]。

在室温下，这些电离能与热能相当（即 0.026eV），因此所有施主（浓度 N_D）都被认为是电离的。n 型半导体中的电荷中性条件表示为

$$N_D + p = n \tag{2.8}$$

将式（2.5）与式（2.8）结合得出以下等式：

$$n^2 + N_D n - n_i^2 = 0 \tag{2.9}$$

当 $N_D \gg n_i$ 时，式（2.9）的解为

$$n = [-N_D + (N_D^2 + 4n_i^2)^{0.5}]/2 \approx N_D \tag{2.10}$$

2.4.2 p 型掺杂

GaN 中受主杂质常用镁，而 4H-SiC 常用铝或硼。与施主能级的位置不同，受主能级 E_A 距离 E_V 要远一些（见图 2.11）。镁在 GaN 中的电离能（E_A-E_V）约为 0.2eV[55]，铝和硼在 4H-SiC 中的电离能分别为 0.2eV 和 0.3eV[56]。这些值远大于实际最高温度下的热能（例如，250℃时为 0.045eV），所以并非所有受主都发生电离。电离受主的浓度由文献[57]给出：

$$N_A^- = N_A/\{1 + 4\exp[(E_A-E_F)/kT]\} \tag{2.11}$$

式中，N_A 为受主浓度；E_F 为费米能级（即处于热动力学平衡状态下，电子能级被占据的概率为 50%）。

电子和空穴浓度[57]分别表示为

$$n = N_C \exp[-(E_C-E_F)/kT] \tag{2.12a}$$

$$p = N_V \exp[-(E_F-E_V)/kT] \tag{2.12b}$$

p 型半导体中的电荷中性条件表示为

$$\begin{aligned} N_V \exp[-(E_F-E_V)/kT] &= N_A/\{1+4\exp[(E_A-E_F)/kT]\} + \\ &\quad N_C \exp[-(E_C-E_F)/kT] \\ &\approx N_A/\{1+4\exp[(E_A-E_F)/kT]\} \end{aligned} \tag{2.13}$$

根据式（2.13），可以通过图形方式确定 E_F，如图 2.15 所示。注意，受主电离比 r_A，即 N_A^-/N_A 不仅取决于 T，还取决于 N_A（例如图 2.15b 中，$T=150$℃，当 $N_A=1\times10^{19}\text{cm}^{-3}$ 时，$r_A=8.6\%$；当 $N_A=1\times10^{17}\text{cm}^{-3}$ 时，$r_A=58\%$）。

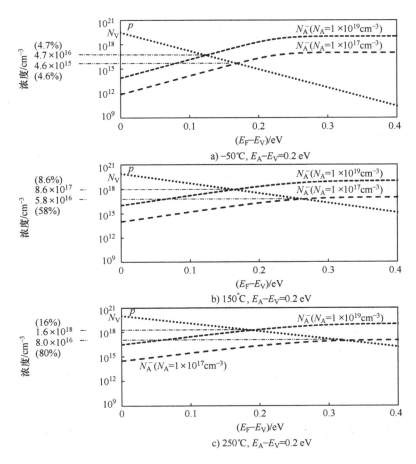

图 2.15 利用 Shockley 图确定 p 型 GaN 中受主电离能为 0.2eV 时,a)−50℃,b)150℃,c)250℃时的费米能级 E_F 和空穴浓度 p。每个图包含了两个受主浓度 N_A

2.5 载流子迁移率

漂移速度 v_{drift} 定义为外加电场 E 时的净载流子速度。在没有与晶格发生碰撞时,电子在电场方向上被加速。当发生碰撞时,电子与晶格交换能量。只要 $|E|$ 很小,交换的能量就很小,晶格不会明显受热。如果基本电荷用 q 表示(1.6×10^{-19}C),则电子受到的力为 $-qE$,获得的动量为 $m_n v_{drift}$,平均散射时间 τ 满足式(2.14),即

$$-qE\tau = m_n v_{drift} \tag{2.14}$$

其中 v_{drift} 为

$$v_{\text{drift}} = -\mu_n E \quad (2.15a)$$
$$\mu_n = q\tau/m_n \quad (2.15b)$$

式中，μ_n为电子迁移率。

由于类似的结论适用于空穴，因此总漂移电流密度J_{drift}可表示为

$$J_{\text{drift}} = (-q)n(-\mu_n E) + qp\mu_p E = q(n\mu_n + p\mu_p)E \quad (2.16)$$

图 2.16 所示为室温下 GaN 和 4H-SiC 中（垂直于 c 轴）μ_n 和 μ_p 与掺杂浓度的关系。注意，μ_n 的值远远大于 μ_p。因此，在低导通电阻的 GaN 和 4H-SiC 功率器件中通常选择电子作为载流子（见第 8 章和第 9 章）。

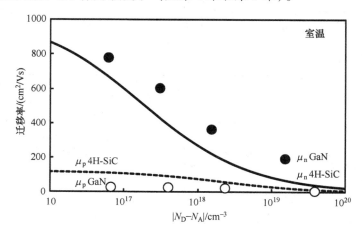

图 2.16 室温下 GaN[55,58] 和 4H-SiC[59,60] 中垂直于 c 轴的电子和空穴迁移率

掺杂浓度 $N_D = 1 \times 10^{16} \text{cm}^{-3}$，在 6H-SiC 中，平行于 c 轴的 μ_n 为 60cm²/Vs，而垂直于 c 轴的 μ_n 为 400cm²/Vs。在 4H-SiC 中，平行于 c 轴的 μ_n 为 900cm²/Vs，垂直于 c 轴的 μ_n 为 800cm²/Vs[61]。这种各向同性的电子输运使 4H-SiC 比 6H-SiC 更适合用作功率器件材料。

2.6 碰撞电离

碰撞电离是由载流子（主要是 GaN[62] 和 4H-SiC[63] 中的空穴）产生的电子-空穴对从电场获得足够能量的过程（见图 2.17），这个过程导致雪崩击穿的发生，其特征在于碰撞电离系数，碰撞电离系数定义为载流子沿电场方向移动单位长度时，耗尽区所产生的电子-空穴对数量。尽管通过许多实验确定了碰撞电离系数[62-68]，但它们之间存在较大的差异，这种差异可能是由缺陷、不均匀电

场以及边缘击穿造成的[62]。本书中使用了报道的最大临界电场（E_critical）（GaN 为 3.75 MV/cm[69] 和 4H-SiC[68] 为 2.50 MV/cm[68]）代替了碰撞电离系数。因此，当最大电场等于 E_critical 时，可以简单地假定雪崩击穿发生。例如，在 4H-SiC 中 E_critical 对 T 和 N_D 的依赖性[68]将被忽略。

$$E_\text{critical} = (2.653 + 2.222 \times 10^{-6} T^2 - 1.166 \times 10^{-3} T) / [1 - 0.25 \log_{10}(N_\text{D}/10^{16})] \text{MV/cm} \quad (2.17)$$

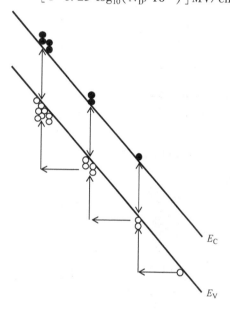

图 2.17　由于空穴的碰撞电离导致的电子和空穴倍增

2.7　品质因数

作为本章最后一个器件物理特性，Baliga 的功率器件品质因数（BFOM）[70]和 Baliga 的高频器件品质因数（BHFFOM）[71]是通过均匀掺杂的平面结漂移区得到的。泊松方程如下

$$\mathrm{d}E(x)/\mathrm{d}x = qN_\text{D}/\varepsilon_\text{r}\varepsilon_0 \quad (2.18)$$

式中，ε_r 为相对介电常数（平行于 GaN[72] 的 c 轴为 10.4，平行于 6H-SiC[73] 的 c 轴为 10.0）；ε_0 为真空中的介电常数（8.85×10^{-14} F/cm）。如图 2.18 所示，式（2.18）的解给出了三角形电场分布。

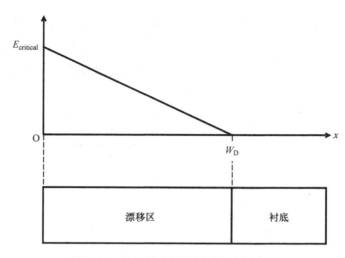

图 2.18 均匀掺杂漂移层及其电场分布

雪崩击穿开始时，$E(0)$ 可以假定为临界击穿电场。电场变为零的漂移区最小厚度为 W_D，因此 N_D 为

$$N_D = (\varepsilon_r \varepsilon_0 / qW_D) E_{critical} \tag{2.19}$$

击穿电压为三角形的面积（即 $E_{critical} W_D / 2$），因此 W_D 为

$$W_D = 2BV / E_{critical} \tag{2.20}$$

对于 n 型漂移区，J_{drift} 可以从式（2.16）近似为

$$J_{drift} \approx qN_D \mu_n (BV / W_D) \tag{2.21}$$

由式（2.19）~式（2.21）获得理想漂移区的特征导通电阻：

$$R_{ideal} = 4BV^2 / BFOM \tag{2.22}$$

其中，

$$BFOM = \varepsilon_r \varepsilon_0 \mu_n E_{critical}^3 \tag{2.23}$$

对于栅极偏压为 V_G 的金属绝缘半导体场效应晶体管（见 9.7 节），单位面积输入电容 C_I（见 9.2.2 节）如下所示[71]

$$C_I = (\varepsilon_r \varepsilon_0 qN_D / 2V_G)^{0.5} \tag{2.24}$$

将式（2.19）和式（2.20）代入式（2.24），可以得到

$$C_I = \varepsilon_r \varepsilon_0 E_{critical} / 2(V_G BV)^{0.5} \tag{2.25}$$

BHFFOM 的定义[71]与频率相关，

$$BHFFOM = 1 / R_{ideal} C_I \tag{2.26}$$

根据式（2.22）~式（2.24）和式（2.26）并结合材料参数得到以下关系：

$$\mathrm{BHFFOM} = \mu_n E_{\text{critical}}^2 V_G^{0.5} / 2BV^{1.5} \quad (2.27)$$

虽然 GaN 与 4H-SiC 的 ε_r、μ_n 和硅相似，分别为 11.7[74] 和 1000cm²/Vs[75]，但 GaN 和 4H-SiC 的临界击穿电场比硅高 10 倍左右，即 0.2MV/cm[75]。如图 2.19 所示，计算得到的 GaN 和 4H-SiC 的 BFOM 和 BHFFOM 分别比硅高出 1000 倍和 100 倍以上。

半导体材料	E_{critical}	ε_r	μ_n	BFOM	BHFFOM
Si	1	1	1	1	1
GaN	19	0.9	0.9	5300	320
4H-SiC	13	0.9	0.9	1500	140

图 2.19 材料属性以及 BFOM 和 BHFFOM 值

2.8 总结

本章介绍了 GaN 和 SiC 的发展历史和物理性质，重点介绍了 GaN 和 SiC 的晶体结构，12.6 节将用到本章对晶体缺陷的描述，尤其是 4H-SiC 中的层错。本章还简要介绍了能带和杂质掺杂。为了引入 BFOM 和 BHFFOM，比较了 GaN 和 4H-SiC 材料的基本特性，如相对介电常数、电子迁移率和临界电场。基于这些特性，本书将比较垂直型 GaN 和 4H-SiC 功率器件的特性。

参 考 文 献

[1] Maruska, H. P., and J. J. Tietjen, "The Preparation and Properties of Vapor-Deposited Single-Crystalline GaN," *Applied Physics Letters*, Vol. 15, No. 10, 1969, pp. 327–329.

[2] Trew, R. J., J.-B. Yan, and P. M. Mock, "The Potential of Diamond and SiC Electronic Devices for Microwave and Millimeter-Wave Power Applications," *Proceedings of the IEEE*, Vol. 79, No. 5, 1991, pp. 598–620.

[3] Amano, H., et al., "Metalorganic Vapor Phase Epitaxial Growth of a High Quality GaN Film Using an AlN Buffer Layer," *Applied Physics Letters*, Vol. 48, No. 5, 1986, pp. 353–355.

[4] Akasaki, I., et al., "Effects of AlN Buffer Layer on Crystallographic Structure and on Electrical and Optical Properties of GaN and $Ga_{1-x}Al_xN$ (0 < x ≤ 0.4) Films Grown on Sapphire Substrate by MOVPE," *Journal of Crystal Growth*, Vol. 98, 1989, pp. 209–219.

[5] Amano, H., et al., "Electron Beam Effects on Blue Luminescence of Zinc-Doped GaN," *Journal of Luminescence*, Vol. 40–41, 1988, pp. 121–122.

[6] Nakamura, S., T. Mukai, and M. Senoh, "High-Power GaN p-n Junction Blue-Light-Emitting Diodes," *Japanese Journal of Applied Physics*, Vol. 30, No. 12A, 1991, pp. L1998–L2001.

[7] Khan, M. A., et al., "High Electron Mobility Transistor Based on a GaN-Al$_x$Ga$_{1-x}$N Heterojunction," *Applied Physics Letters*, Vol. 63, No. 9, 1993, pp. 1214–1215.

[8] Pearton, S. J., C. R. Abernathy, and F. Ren, *Gallium Nitride Processing for Electronics, Sensors, and Spintronics*, London: Springer, 2006, p. 6.

[9] Kelly, M. K., et al., "Large Freestanding GaN Substrates by Hydride Vapor Phase Epitaxy and Laser-Induced Liftoff," *Japanese Journal of Applied Physics*, Vol. 38, No. 3A, 1999, pp. L217–L219.

[10] Johnson, J. W., et al., "Breakdown Voltage and Reverse Recovery Characteristics of Freestanding GaN Schottky Rectifiers," *IEEE Transactions on Electron Devices*, Vol. 49, No. 1, 2002, pp. 32–36.

[11] Suda, J., and M. Horita, "Characterization of n-Type and p-Type GaN Layers Grown on Freestanding GaN Substrates," *Compound Semiconductor Week*, Toyama, Japan, June 26–30, 2016, paper WeB1-1.

[12] Tairov, Y. M., and V. F. Tsvetkov, "Investigation of Growth Processes of Ingots of Silicon Carbide Single Crystals," *Journal of Crystal Growth*, Vol. 43, 1978, pp. 209–212.

[13] Scace, R. I., and G. A. Slack, "Solubility of Carbon in Silicon and Germanium," *Journal of Chemical Physics*, Vol. 30, No. 6, 1959, pp. 1551–1555.

[14] Wesch, W., "Silicon Carbide: Synthsis and Processing," *Nuclear Instruments and Methods in Physics Research B*, Vol. 116, 1996, pp. 305–321.

[15] Kuroda, N., et al., "Step-Controlled VPE Growth of SiC Single Crystals at Low Temperatures," *Solid State Devices and Materials*, Tokyo, Aug. 25–27, 1987, pp. 227–230.

[16] Matsunami, H., "Technological Breakthroughs in Growth Control of Silicon Carbide for High Power Electronic Devices," *Japanese Journal of Applied Physics*, Vol. 43, No. 10, 2004, pp. 6835–6847.

[17] Ben-Yaacov, I., et al., "AlGaN/GaN Current Aperture Vertical Electron Transistors with Regrown Channels," *Journal of Applied Physics*, Vol. 95, No. 4, 2004, pp. 2073–2078.

[18] Kanechika, M., et al., "A Vertical Insulated Gate AlGaN/GaN Hetrojunction field Effect Transistor," *Japanese Journal of Applied Physics*, Vol. 46, No. 21, 2007, pp. L503–L505.

[19] Shibata, D., et al., "1.7 kV/1.0 mΩcm^2 Normally-Off Vertical GaN Transistor on GaN Substrate with Regrown p-GaN/AlGaN/GaN Semipolar Gate Structure," *International Electron Devices Meeting*, San Francisco, Dec. 5–7, 2016, pp. 10.1.1–10.1.4.

[20] Ueno, M., et al., "X-Ray Observation of the Structural Phase Transition of Aluminum Nitride Under High Pressure," *Physical Review B*, Vol. 45, No. 1, 1992, pp. 10123–10126.

[21] Ueno, M., et al., "Stability of the Wurtzite-Type Under High Pressure: GaN and InN," *Physical Review B*, Vol. 49, No. 17, 1994, pp. 14–21.

[22] Yeh, C.-Y., et al., "Zinc-Blende–Wurtzite Polytypism in Semiconductors," *Physical Review B*, Vol. 46, No. 16, 1992, pp. 10086–10097.

[23] Frank, F. C., "On Miller-Bravais Indices and Four-Dimensional Vectors," *Acta Crystallographica*, Vol. 18, 1965, pp. 862–866.

[24] Li, X., et al., "Properties of GaN Layers Grown on N-Face Freestanding GaN Substrates," *Journal of Crystal Growth*, Vol. 413, 2015, pp. 81–85.

[25] Kojima, K., et al., "Low-Resistivity m-Plane Freestanding GaN Substrate with Very Low Point-Defect Concentrations Grown by Hydride Vapor Phase Epitaxy on a GaN Seed Crystal Synthesized by the Ammonothermal Method," *Applied Physics Express*, Vol. 8, 2015, pp. 095501-1–095501-4.

[26] Wu, Y.-H., et al., "Freestanding a-Plane GaN Substrates Grown by HVPE," *SPIE Proceedings*, Vol. 8262, 2012, pp. 82621Z-1–82621Z-5.

[27] Yazdi, G. R., et al., "Freestanding AlN Single Crystals Enabled by Self-Organization of 2H-SiC Pyramids on 4H-SiC Substrates," *Applied Physics Letters*, Vol. 94, No. 8, 2009, pp. 082109-1–082109-3.

[28] Satoh, I., et al., "Development of Aluminum Nitride Single-Crystal Substrates," *SEI Technical Review*, Vol. 71, 2010, pp. 78–82.

[29] Novikov, S. V., et al., "Molecular Beam Epitaxy of Freestanding Bulk Wurtzite $Al_xGa_{1-x}N$ layers Using a Highly Efficient RF Plasma Source," *Physica Status Solidi C*, Vol. 13, No. 5–6, 2016, pp. 217–220.

[30] Yoo, W. S., and H. Matsunami, "Solid-State Phase Transformation in Cubic Silicon Carbide," *Japanese Journal of Applied Physics*, Vol. 30, No. 3, 1991, pp. 545–553.

[31] Kittel, C., *Introduction to Solid State Physics* (Eighth Edition), New Jersey: John Wiley & Sons, 2005, p.188 and pp. 585–586.

[32] Bockstedte, M., et al., "Ab Initio Study of the Migration of Intrinsic Defects in 3C-SiC," *Physical Review B*, Vol. 68, No. 20, 2003, pp. 205201-1–205201-17.

[33] Bracht, H., "Self- and Foreign-Atom Diffusion in Semiconductor Isotope Heterostructures. I. Continuum Theoretical Calculations," *Physical Review B*, Vol. 75, No. 3, 2007, pp. 035210-1–035210-3.

[34] https://www.silvaco.com/products/tcad/process_simulation/process_simulation.html.

[35] Tokuda, Y., et al., "DLTS Study of n-Type GaN Grown by MOCVD on GaN Substrates," *Superlattices and Microstructures*, Vol. 40, 2006, pp. 268–273.

[36] Dalibor, T., et al., "Deep Defect Centers in Silicon Carbide Monitored with Deep Level Transient Spectroscopy," *Physica Status Solidi A*, Vol. 162, No. 1, 1997, pp. 199–225.

[37] Hemmingsson, C., et al., "Deep Level Defects in Electron-Irradiated 4H-SiC Epitaxial Layers," *Journal of Applied Physics*, Vol. 81, No. 9, 2007, pp. 6155–6159.

[38] Kimoto, T., and J. A. Cooper, *Fundamentals of Silicon Carbide Technology*, Singapore: John Wiley & Sons, 2014, pp. 161–172.

[39] Evoy, S., et al., "Scanning Tunneling Microscope-Induced Luminescence of GaN at Threading Dislocations," *Journal of Vacuum Science and Technology B*, Vol. 17, No. 1, 1999, pp. 29–32.

[40] Northrop, J. E., "Theory of ($12\bar{1}0$) Prismatic Stacking Fault in GaN," *Applied Physics Letters*, Vol. 72, No. 18, 1998, pp. 2316–2318.

[41] Stampfl, C., and C. G. Van de Walle, "Energetics and Electronic Structure of Stacking Faults in AlN, GaN, and InN," *Physical Review B*, Vol. 57, No. 24, 1998, pp. R15052–R15055.

[42] Hong, M. H., et al., "Stacking Fault Energy of 6H-SiC and 4H-SiC Single Crystals," *Phylosophical Magazine A*, Vol. 80, No. 4, 2000, pp. 919–935.

[43] Daudin, B., J. L. Rouviere, and M. Arlery, "Polarity Determination of GaN Films by Ion Channeling and Convergent Beam Electron Diffraction," *Applied Physics Letters*, Vol. 69, No. 17, 1996, pp. 2480–2482.

[44] Beach, R. A., and T. C. McGill, "Piezoelectric Fields in Nitride Devices," *Journal of Vacuum Science and Technology B*, Vol. 17, No. 4, 1999, pp. 1753–1756.

[45] Kamata, T., et al., "Structural Transformation of Screw Dislocations via Thick 4H-SiC Epitaxial Growth," *Japanese Journal of Applied Physics*, Vol. 39, No. 12A, 2000, pp. 6496–6500.

[46] Kamata, T., et al., "Influence of 4H-SiC Growth Conditions on Micropipes Dissociation," *Japanese Journal of Applied Physics*, Vol. 41, No. 10B, 2002, pp. L1137–L1139.

[47] Perry, P. B., and R. F. Rutz, "The optical Absorption Edge of Single-Crystal AlN Prepared by a Close-Spaced Vapor Process," *Applied Physics Letters*, Vol. 33, No. 4, 1978, pp. 319–321.

[48] Muth, J. F., et al., "Absorption Coefficient, Energy Gap, Exciton Binding Energy, and Recombination Lifetime of GaN Obtained from Transmission Measurements," *Applied Physics Letters*, Vol. 71, No. 18, 1997, pp. 2572–2574.

[49] Sridhara, S. G., et al., "Absorption Coefficient of 4H Silicon Carbide from 3900 to 3250 Angstrom," *Journal of Applied Physics*, Vol. 84, No. 5, 1998, pp. 2963–2964.

[50] Monemar, B., et al., "Recombination of Free and Bound Excitons in GaN," *Physica Status Solidi B*, Vol. 245, No. 9, 2008, pp. 1723–1740.

[51] Bougrov, V., et al., in *Properties of Advanced Semiconductor Materials GaN, AlN, InN, BN, SiC, SiGe* (eds. Levinshtein, M. E., Rumyantsev, S. L., and Shur, M. S.), New York: John Wiley & Sons, 2001, pp. 1–30.

[52] http://www.ioffe.ru/SVA/NSM/Semicond/GaN/bandstr.html.

[53] http://www.ioffe.ru/SVA/NSM/Semicond/SiC/bandstr.html.

[54] Choyke, W. J., and G. Pensl, "Physical Properties of SiC," *MRS Bulletin*, Vol. 22, No. 3, 1997, pp. 25–29.

[55] Horita, M., et al., "Hall-Effect Measurements of Metalorganic Vapor-Phase Epitaxy-Grown p-Type Homoepitaxial GaN Layers with Various Mg Concentrations," *Japanese Journal of Applied Physics*, Vol. 56, 2017, pp. 031001-1-031001-4.

[56] Heera, V., D. Panknin, and W. Skorupa, "P-Type Doping of SiC by High Dose Al Implantation–Problems and Progress," *Applied Surface Science*, Vol. 184, No. 1–4, 2001, pp. 307–316.

[57] Sze, S. M., and K. K. Ng, *Physics of Semiconductor Devices* (Third Edition), New Jersey: John Wiley & Sons, 2007, pp. 17–27.

[58] Kyle, E. C. H., et al., "High-Electron-Mobility GaN Grown on Freestanding GaN Templates by Ammonia-Based Molecular Beam Epitaxy," *Journal of Applied Physics*, Vol. 115, 2014, pp. 193702-1–193702-12.

[59] Kagamihara, S., et al., "Parameters Required to Simulate Electric Characteristics," *Journal of Applied Physics*, Vol. 96, No. 10, 2004, pp. 5601–5606.

[60] Matsuura, H., et al., "Dependence of Acceptor Levels and Hole Mobility on Acceptor Density and Temperature in Al-Doped p-Type 4H-SiC Epilayers," *Journal of Applied Physics*, Vol. 96, No. 5, 2004, pp. 2708–2715.

[61] http://www.iue.tuwien.ac.at/phd/ayalew/node21.html.

[62] Baliga, B. J., "Gallium Nitride Devices for Power Electric Applications," *Semiconductor Science and Technology*, Vol. 28, 2013, pp. 1–8.

[63] Konstantinov, A. O., et al., "Ionization Rates and Critical Fields in 4H Silicon Carbide," *Applied Physics Letters*, Vol. 71, No. 1, 1997, pp. 90–92.

[64] Raghunathan, R., and B. J. Baliga, "Temperature Dependence of Hole Impact Ionization Coefficients in 4H and 6H SiC," *Solid State Electronics*, Vol. 43, No. 2, 1999, pp. 199–211.

[65] Hatakeyama, T., et al., "Impact Ionization Coefficients of 4H Silicon Carbide," *Applied Physics Letters*, Vol. 85, No. 8, 2004, pp. 1380–1382.

[66] Loh, W. S., et al., "Impact Ionization Coefficients in 4H-SiC," *IEEE Transactions on Electron Devices*, Vol. 55, No. 8, 2008, pp. 1984–1990.

[67] Green, J. E., et al., "Impact Ionization Coefficients in 4H-SiC by Ultralow Excess Noise Measurements," *IEEE Transactions on Electron Devices*, Vol. 59, No. 4, 2012, pp. 1030–1036.

[68] Niwa, H., J. Suda, and T. Kimoto, "Impact Ionization Coefficients in 4H-SiC Toward Ultrahigh-Voltage Power Devices," *IEEE Transactions on Electron Devices*, Vol. 62, No. 10, 2015, pp. 3326–3333.

[69] Ozbek, A. M., and B. J. Baliga, "Planar Nearly Ideal Edge-Termination Technique for GaN Devices," *IEEE Electron Device Letters*, Vol. 32, No. 3, 2011, pp. 300–302.

[70] Baliga, B. J., "Semiconductors for High-Voltage, Vertical Channel FETs," *Journal of Applied Physics*, Vol. 53, No. 3, 1982, pp. 1759–1764.

[71] Baliga, B. J., "Power Semiconductor Device Figure of Merit for High-Frequency Applications," *IEEE Electron Device Letters*, Vol. 10, No. 10, 1989, pp. 455–457.

[72] Barker, Jr., A. S., and M. Ilegems, "Infrared Lattice Vibrations and Free-Electron Dispersion in GaN," *Physical Review B*, Vol. 7, No. 2, 1973, pp. 743–750.

[73] Patrick, L., and W. J. Choyke, "Static Dielectric Constant of SiC," *Physical Review B*, Vol. 2, No. 6, 1970, pp. 2255–2256.

[74] http://www.ioffe.ru/SVA/NSM/Semicond/Si/basic.html.

[75] Baliga, B. J., *Fundamentals of Power Semiconductor Devices*, New York: Springer Science, 2008, p. 16.

第 3 章

p-n 结

3.1 引言

p-n 结在双极功率器件（见第 10 章）和结终端（见第 11 章）中都起到了非常重要的作用。本章将首先介绍扩散电流方程，然后推导连续性方程，并将比较 GaN、4H-SiC 和 Si 的载流子复合寿命。然后将解释 Shockley[1] 提出的电流/电压特性基本理论。此外，本章还将讨论多维效应对非自对准台面型 p-n 结电流/电压特性的影响。

由于 GaN 和 4H-SiC 的电子迁移率高于其空穴迁移率（请参见 2.5 节），这类功率器件通常由 n 型衬底材料制成。因此，本章将以 n 型 GaN 和 4H-SiC 为主，同时分析其与重掺杂 p 型（p^+）GaN 和 4H-SiC 形成的结。

3.2 扩散

当载流子浓度具有梯度时，载流子通过扩散从高浓度区域迁移到低浓度区域。由菲克定律[2]，电子和空穴的一维扩散电流密度可以分别表示为

$$J_{n\_diffusion} = qD_n(\partial n/\partial x) \tag{3.1a}$$

和

$$J_{p\_diffusion} = -qD_p(\partial p/\partial x) \tag{3.1b}$$

式中，D_n 和 D_p 分别为电子和空穴的扩散系数。在既不重掺杂（即未退化）也不存在外部施加电场的 n 型半导体中，电子的漂移电流密度（由式（2.16）的 $qn\mu_n E$ 给出）平衡了 $J_{n\_diffusion}$，如下所示

$$qn\mu_n E + qD_n(\partial n/\partial x) = 0 \tag{3.2}$$

热平衡条件下,费米能级 E_F 恒定。因此,关于 x 方向的电子浓度 $n = N_C \exp[-(E_C-E_F)/kT]$ 的偏导数(请参阅式(2.12a))变为

$$\partial n/\partial x = -(N_C/kT)(\partial E_C/\partial x)\exp[-(E_C-E_F)/kT] \tag{3.3}$$

同时

$$qE = \partial E_C/\partial x \tag{3.4}$$

式(3.3)变为

$$\partial n/\partial x = -qEn/kT \tag{3.5}$$

结合式(3.2)和式(3.5)得出 n 型半导体的爱因斯坦关系如下所示

$$D_n = (kT/q)\mu_n \tag{3.6a}$$

同样地,对于非简并的 p 型半导体,爱因斯坦关系为

$$D_p = (kT/q)\mu_p \tag{3.6b}$$

3.3 连续性方程

取 x 处无穷小晶体截面(厚度为 dx;横截面积为 A)进行一维处理(见图 3.1)。流入截面和流出截面的电子数分别为 $J_n(x)/(-q)$ 和 $J_n(x+dx)/(-q)$。电子单位体积的产生率和复合率分别由 G_n 和 R_n 表示,截面电子数的变化速率由式(3.7)给出:

$$(\partial n/\partial t)Adx = [J_n(x)/(-q) - J_n(x+dx)/(-q)]A + (G_n - R_n)Adx \tag{3.7}$$

将泰勒级数右边的第二项扩展为以下连续性方程:

$$\partial n/\partial t = (1/q)[\partial J_n(x)/\partial x] + (G_n - R_n) \tag{3.8a}$$

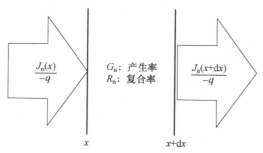

图 3.1 无限小截面电子浓度的增加(厚度为 dx;横截面积为 A)与电子流入截面的净流量以及复合产生的过剩电子关系

对于空穴而言，连续性方程可由式（3.8b）获得，将式（3.8a）右侧用（$-q$）的符号替换：

$$\partial p/\partial t = (-1/q)[\partial J_p(x)/\partial x] + (G_p - R_p) \quad (3.8b)$$

根据式（2.16）（即 $J_{drift} = q(n\mu_n + p\mu_p)E$），电子和空穴漂移电流密度为

$$J_{n\_drift} = qn\mu_n E \quad (3.9a)$$

和

$$J_{p\_drift} = qp\mu_p E \quad (3.9b)$$

将式（3.9a）和式（3.9b）分别与式（3.1a）和式（3.1b）结合，得到电子和空穴的总电流密度方程如下

$$J_n = qn\mu_n E + qD_n(\partial n/\partial x) \quad (3.10a)$$

和

$$J_p = qp\mu_p E - qD_p(\partial p/\partial x) \quad (3.10b)$$

将式（3.10a）和式（3.10b）分别代入式（3.8a）和式（3.8b）中，并假设迁移率和扩散系数与 x 无关，分别给出电子和空穴的连续性方程最终形式如下所示

$$\partial n/\partial t = \mu_n n(x)(\partial E/\partial x) + \mu_n E(x)(\partial n/\partial x) + D_n \partial^2 n/\partial x^2 + (G_n - R_n) \quad (3.11a)$$

和

$$\partial p/\partial t = -\mu_p p(x)(\partial E/\partial x) - \mu_p E(x)(\partial p/\partial x) + D_p \partial^2 p/\partial x^2 + (G_p - R_p) \quad (3.11b)$$

3.4 载流子复合寿命

诸如光子或热能之类的外部激励会产生过剩的载流子，当消除外部激励时，过剩载流子浓度主要通过带间复合（见图3.2a）、间接复合（见图3.2b）和俄歇复合（见图3.2c）的方式衰减达到平衡值。

3.4.1 带间复合寿命

带间复合率 $R_{band-to-band}$ 与电子和空穴浓度成正比，即

$$R_{band-to-band} = Bnp \quad (3.12)$$

式中，B 为带间复合系数（GaN 为 $1.2×10^{-11} cm^3/s$[3]，4H-SiC 为 $1.5×10^{-12} cm^3/s$[4]，而 Si 为 $4.7×10^{-15} cm^3/s$[5]）。由于直接带隙导致 GaN 拥有较大的 B（请参见2.3节）。对间接带隙半导体，4H-SiC 的 B 是 Si 的300倍以上，B 的这一巨大差异将在3.4.4节中进行定量讨论。

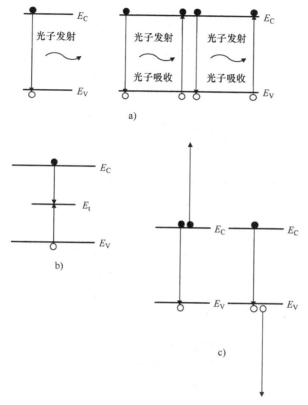

图 3.2 a) 带间复合，b) 间接复合和 c) 俄歇
复合示意图。图 a) 中还显示了辐射导致带间
复合的再吸收作用（即光子回收）

在大注入条件下，即当过剩载流子浓度 p（$p=n$ 为电中性条件）大于多数载流子浓度时，带间复合对于载流子寿命确实具有重要影响。分别给出大注入条件下 n 型半导体中的电子和空穴浓度，如下所示

$$n_n = N_D + p \approx p \qquad (3.13a)$$

和

$$p_n = p_{no} + p \approx p \qquad (3.13b)$$

式中，p_{no} 为热平衡下的空穴浓度。

因此，在大注入条件下，n 型半导体中的 $R_{\text{band-to-band}}$ 为

$$R_{\text{band-to-band\_p}} = B\Delta p^2 \equiv \Delta p / \tau_{\text{band-to-band\_p}} \qquad (3.14a)$$

大注入条件下的带间复合寿命为

$$\tau_{\text{band-to-band\_p}} = 1/(B\Delta p) \qquad (3.14b)$$

如 2.3 节所述，GaN 的光吸收系数比 4H-SiC 大得多。因此，对于 GaN 而言，辐射复合通常会被重新吸收（即光子回收）（见图 3.2a），从而有效地增加了 $\tau_{\text{band-to-band\_p}}$，也减小了 B 的值。第 4 章将详细描述光子回收效应。

3.4.2 间接复合寿命

在任何半导体中，由于点缺陷（见 2.2.3 节）和杂质原子等因素，实际上能隙中还存在局域态（见图 3.2b），根据 Shockley、Read[6] 和 Hall[7] 的分析，稳态下单个深能级复合中心的复合率 U_{SRH} 为

$$U_{\text{SRH}} = \sigma_n \sigma_p v_{\text{th}} N_t (np - n_i^2) / [\sigma_n(n+n_1) + \sigma_p(p+p_1)] \quad (3.15)$$
$$= (np - n_i^2) / [(n+n_1)/(\sigma_p v_{\text{th}} N_t) + (p+p_1)/(\sigma_n v_{\text{th}} N_t)]$$

式中，σ_n 和 σ_p 分别为电子和空穴的俘获截面；v_{th} 为热速度；N_t 为阱浓度。

当 E_F 与阱能级 E_t 一致时，n_1 和 p_1 分别为平衡状态下的电子和空穴浓度，如下

$$n_1 = n_i \exp[(E_t - E_i)/kT] \quad (3.16a)$$

和

$$p_1 = n_i \exp[(E_i - E_t)/kT] \quad (3.16b)$$

式中，E_i 为本征费米能级（即中间带隙能级）。

从而，重掺杂下 p 型和 n 型半导体中的少数载流子寿命分别定义为

$$\tau_{p0} = 1/(\sigma_p v_{\text{th}} N_t) \quad (3.17a)$$

和

$$\tau_{n0} = 1/(\sigma_n v_{\text{th}} N_t) \quad (3.17b)$$

式（3.15）可以表示为

$$U_{\text{SRH}} = (np - n_i^2)/[\tau_{p0}(n+n_1) + \tau_{n0}(p+p_1)] \quad (3.18)$$

假设 $\sigma_n = \sigma_p \equiv \sigma_0$ 和 $\tau_{n0} = \tau_{p0} \equiv \tau_0$，则式（3.18）中包含 E_t 的 U_{SRH} 可以表示为

$$U_{\text{SRH}} = (np - n_i^2)/\{n + p + 2n_i \cosh[(E_t - E_i)/kT]\tau_0\} \quad (3.19)$$

此处无电流通过的 n 型半导体会受到相等数量过剩电子和空穴的干扰。如果是小注入条件，则 Δn 和 Δp 远小于 $(n_0 + p_0)$，其中 n_0 和 p_0 分别表示热平衡状态下的电子和空穴浓度，过剩空穴的连续性方程（3.11b）变为

$$\partial \Delta p / \partial t = G_p - R_p = -U_{\text{SRH}} \quad (3.20)$$
$$= [-(n_0 + p_0)\Delta p] / \{n_0 + p_0 + 2n_i \cosh[(E_t - E_i)/kT]\tau_0\}$$

对于中间带隙能级附近的复合中心，$(E_t - E_i)$ 可以忽略不计，因此式（3.20）也可以表示为

$$\partial \Delta p / \partial t \equiv -\Delta p / \tau_{\text{SRH\_p}} \tag{3.21a}$$

和

$$\tau_{\text{SRH\_p}} = \tau_0 = \tau_{\text{p0}} = 1/(\sigma_\text{p} v_{\text{th}} N_\text{t}) \tag{3.21b}$$

式（3.21b）表明，小注入条件下 n 型半导体中空穴的 SRH 复合寿命（$\tau_{\text{SRH\_p}}$）与多数载流子浓度 N_D 无关。

3.4.3 俄歇复合寿命

当过剩载流子在掺杂浓度高的区域复合时，则不能忽略电子与空穴之间的直接复合（见图 3.2c）。在这种复合过程中（称为俄歇复合）3 个自由载流子相互作用，即两个电子和一个空穴或两个空穴和一个电子（见图 3.2c）。俄歇复合率 U_A 为

$$U_\text{A} = C_\text{n} n (np - n_\text{i}^2) + C_\text{p} p (np - n_\text{i}^2) \tag{3.22}$$

式中，C_n 和 C_p 分别为过剩载流子电子或空穴的相互作用系数。

n 型半导体在俄歇复合的情况下，如果 $C_\text{n} \approx C_\text{p}$，则式（3.22）右侧的第二项小于式（3.22）右侧的第一项。因此，过剩空穴的俄歇复合寿命 Δp 为

$$\tau_{\text{A\_p}} = \Delta p / U_\text{A} \approx 1/(C_\text{n} N_\text{D}^2) \tag{3.23}$$

式中，GaN 的 C_n 值为 $4.5 \times 10^{-31} \text{cm}^6/\text{s}$[8]，4H-SiC 的 C_n 值为 $5.0 \times 10^{-31} \text{cm}^6/\text{s}$[9]，Si 的 C_n 值为 $2.8 \times 10^{-31} \text{cm}^6/\text{s}$[10]。

3.4.4 载流子复合寿命的整体表达式

如果忽略 Δp 对 $\tau_{\text{SRH\_p}}$ 的影响，n 型半导体中空穴复合寿命的整体表达式为

$$\tau_\text{p} = 1/[(1/\tau_{\text{band-to-band\_p}}) + (1/\tau_{\text{SRH\_p}}) + (1/\tau_{\text{A\_p}})] \tag{3.24}$$

3.4.1 节和 3.4.3 节中描述的 B 和 C_n 值可用于计算 τ_p，当 $N_\text{D} = 1 \times 10^{14} \text{cm}^{-3}$ 时，它是 Δp 的函数（见图 3.3）。对于硅而言，当大注入条件下的 $\tau_{\text{SRH\_p}}$ 超过 $10\mu\text{s}$ 时，俄歇复合会限制 τ_p（见图 3.3a）。另一方面，对于 4H-SiC 而言，当大注入条件下的 $\tau_{\text{SRH\_p}}$ 超过 $1\mu\text{s}$ 时，带间复合会限制 τ_p（见图 3.3b），造成这种结果的主要原因是 4H-SiC 的 B 相对较大。

对于 GaN，如果光子回收的影响不存在，则在一定的过剩载流子浓度下，GaN 的 τ_p 会降低到 4H-SiC τ_p 的 10% 左右。这是因为当 $\tau_{\text{SRH\_p}}$ 为 100ns 时，带间复合会限制 τ_p（见图 3.3c）。然而，在光子回收的作用下，τ_p 对 GaN 过剩载流子浓度的影响与 4H-SiC 相当。例如，在 AlGaAs/GaAs 异质结构中，光

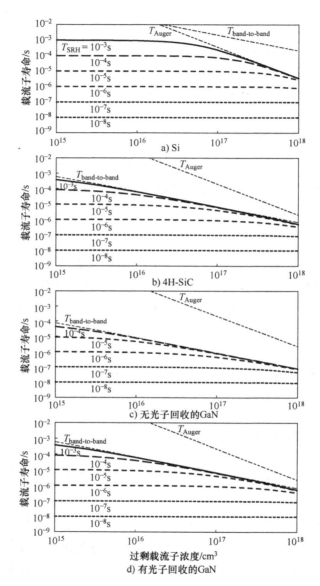

图 3.3 n 型复合情况下过剩载流子浓度的计算曲线 a) Si，b) 4H-SiC，c) 无光子回收的 GaN 和 d) 有光子回收的 GaN（光子回收因子：8）。Si 的带间复合和俄歇复合系数分别为 $4.7 \times 10^{-15} \mathrm{cm^3/s}$ 和 $2.8 \times 10^{-31} \mathrm{cm^6/s}$，4H-SiC 为 $1.5 \times 10^{-12} \mathrm{cm^3/s}$ 和 $5.0 \times 10^{-31} \mathrm{cm^6/s}$，GaN 为 $1.2 \times 10^{-11} \mathrm{cm^3/s}$ 和 $4.5 \times 10^{-31} \mathrm{cm^6/s}$

子回收因子 Φ（即有效带间复合寿命 $\tau_{\text{band-to-band}}^{\text{eff}}$ 除以 $\tau_{\text{band-to-band\_p}}$）的取值为 4 ~ 6[11]。在图 3.3d 中，Φ 的值仅被假设为 8，因为 GaN 的 $\tau_{\text{band-to-band}}^{\text{eff}}$ 等于 4H-SiC 的 $\tau_{\text{band-to-band\_p}}$。由于 GaN 的 C_n 与 4H-SiC 的 C_n 大致相同（请参阅 3.4.3 节），因此 τ_p 对过剩载流子浓度的影响在 GaN 和 4H-SiC 中类似。

3.5 一维 p$^+$n 突变结的耗尽区宽度

首先假定电荷耗尽区在 p$^+$ 区域的宽度为 W_{Dp}，而在 n 区域为 W_{Dn}，可用下面的耗尽层近似做简单的分析（见图 3.4a）。由于中性区的电场几乎为零，因此必满足以下方程：

$$N_A W_{\text{Dp}} = N_D W_{\text{Dn}} \tag{3.25}$$

由于泊松方程表示为

$$\partial E / \partial x = -qN_A / \varepsilon_r \varepsilon_o, \, -W_{\text{Dp}} \leq x \leq 0 \tag{3.26a}$$

和

$$\partial E / \partial x = qN_D / \varepsilon_r \varepsilon_o, \, 0 \leq x \leq W_{\text{Dn}} \tag{3.26b}$$

通过对式（3.26a）和式（3.26b）进行积分可获得如下电场：

$$E(x) = -qN_A(x + W_{\text{Dp}}) / \varepsilon_r \varepsilon_o, \, -W_{\text{Dp}} \leq x \leq 0 \tag{3.27a}$$

和

$$E(x) = -E_{\max} + qN_D x / \varepsilon_r \varepsilon_o = -qN_D(W_{\text{Dn}} - x) / \varepsilon_r \varepsilon_o, \, 0 \leq x \leq W_{\text{Dn}} \tag{3.27b}$$

其中最大电场 $E_{\max}$（见图 3.4b）表示为

$$E_{\max} = qN_A W_{\text{Dp}} / \varepsilon_r \varepsilon_o = qN_D W_{\text{Dn}} / \varepsilon_r \varepsilon_o \tag{3.28}$$

如图 3.4b 和 c 所示，在 p$^+$n 突变结的情况下，W_{Dp} 和 ψ_p 比 W_{Dn} 和 ψ_n 小得多，所以总耗尽区宽度 W_D 和内建电势 ψ_{bi} 可以分别用 W_{Dn} 和 ψ_n 近似估算。对式（3.27b）积分可得 ψ_{bi} 为

$$\psi_{\text{bi}} \approx qN_D W_{\text{Dn}}^2 / 2\varepsilon_r \varepsilon_o \tag{3.29}$$

从式（3.29）可知，一维 p$^+$n 突变结的 W_D 表示为

$$W_D \approx (2\varepsilon_r \varepsilon_o \psi_{\text{bi}} / qN_D)^{0.5} \tag{3.30}$$

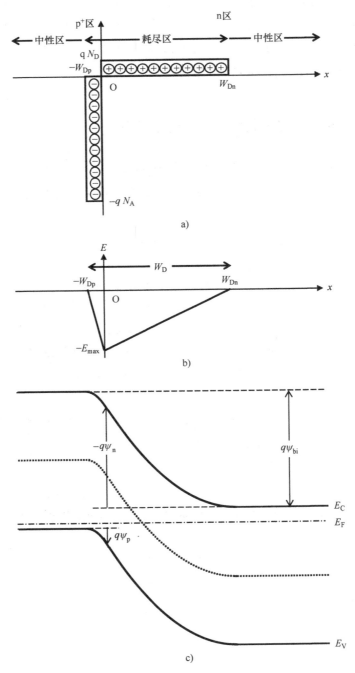

图 3.4 热平衡条件下的 p^+n 突变结：a) 耗尽层近似下的空间电荷分布，b) 电场分布和 c) 能带图

3.6 一维正向电流/电压特性

在一维 p^+n 结上施加一个电压源,其中 n 区接地且 p^+ 区接高电位。假定 p^+n 结没有开启,因此载流子浓度仅受施加电压 V 的影响。此外,假定欧姆电压降被忽略,因此电压 V 完全落在结上,即电压为 $\psi_{bi}-V$。

3.6.1 小注入条件

不论 p^+n 结是处于平衡状态还是处于偏置状态,都可以假定在靠近 p^+n 结的 n 区(即 $x=W_{Dn}\approx W_D$)准中性边界处的电子浓度等于 N_D。因此,将 ψ_{bi} 替换为 $\psi_{bi}-V$,偏置状态下的 W_D 可由式(3.30)得到

$$W_D \approx [2\varepsilon_r\varepsilon_o(\psi_{bi}-V)/qN_D]^{0.5} \tag{3.31}$$

如果将热平衡状态下 p^+ 区的空穴浓度表示为 p_{p0},则小注入条件下的 $p(W_D)$ 可近似为

$$p(W_D) \approx p_{p0}\exp[-q(\psi_{bi}-V)/kT] = (n_i^2/N_D^+)\exp(qV/kT) \tag{3.32}$$

3.6.1.1 扩散电流

稳态下中性 n 区的连续性方程(3.11b)为

$$0 = D_p\partial^2\Delta p/\partial x^2 - \Delta p(x)/\tau_{SRH\_p} \tag{3.33}$$

式(3.33)的简单指数解为

$$\Delta p(x) = A\exp[-(x-W_D)/(D_p\tau_{SRH\_p})^{0.5}] + B\exp[(x-W_D)/(D_p\tau_{SRH\_p})^{0.5}] \tag{3.34}$$

式中,常数 A 和 B 由边界条件确定。

如果中性 n 区宽度远大于扩散长度 $L_p \equiv (D_p\tau_{SRH\_p})^{0.5}$,则 L_p 表示复合之前从 p^+ 区注入的空穴在中性 n 区中移动的平均距离。由于 $\Delta p(x)$ 随 x 的增加而减小,因此式(3.34)中的 B 为零。结合式(3.32)和式(3.34),得出

$$\Delta p(x) = (n_i^2/N_D^+)[\exp(qV/kT)-1]\exp[-(x-W_D)/L_p] \tag{3.35}$$

由于在 $x=W_D$ 处的空穴电流仅存在扩散的形式(见图 3.5a),则 p^+n 结的总电流密度为

$$J_{total} = Jp(x)|_{x=W_D} = -qD_p\partial\Delta p/\partial\Delta x|_{x=W_D} = J_S^{LL}[\exp(qV/kT)-1] \tag{3.36a}$$

和

$$J_S^{LL} = qD_pn_i^2/(N_D+L_p) \tag{3.36b}$$

在 p⁺ 区中 N_A^- 减小的情况下（见图 3.5b），J_{total} 可以表示为 $J_{p\_diffusion}(W_{Dn})$ 和 $J_{n\_diffusion}(-W_{Dp})$ 之和。结合式（3.35）、式（3.36a）和式（3.36b），J_{total} 可以表示为

$$J_{total} = qn_i^2 [D_p/(N_D+L_p) + D_n/(N_A-L_n)][\exp(qV/kT)-1] \quad (3.37)$$

若 p 区的厚度（W_p）小于 L_n（见图 3.5c），则 $J_{n\_diffusion}(-W_{Dp})$ 的表达式变为

$$J_{n\_diffusion}(-W_{Dp}) = qn_i^2 D_n/(N_A-W_p)[\exp(qV/kT)-1] \quad (3.38)$$

因此 J_{total} 为

$$J_{total} = qn_i^2 [D_p/(N_D+L_p) + D_n/(N_A-W_p)][\exp(qV/nkT)-1] \quad (3.39)$$

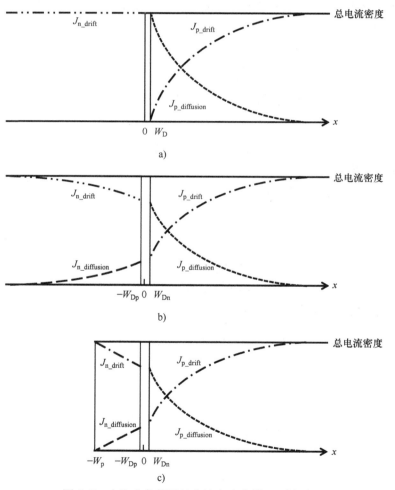

图 3.5 小注入条件下的电流密度分量 a) 厚 p⁺n，
b) 厚 p⁻n 和 c) 薄 p⁻n 突变结

当二极管电流密度与 [exp(qV/nkT)-1] 成正比时，n 称为理想因子。式 (3.36a)、式 (3.37) 和式 (3.39) 表明，小注入条件下理想扩散电流的理想因子均为1。

3.6.1.2 空间电荷复合电流

由式 (3.36a)、式 (3.37) 和式 (3.39) 表示的 J_{total} 还存在空间电荷复合电流密度 $J_{recombination}$。当外加电压 V 很小时，$J_{recombination}$ 会变得比 J_{total} 更大，所以 $J_{recombination}$ 不可忽略。根据质量作用定律和式 (3.32) 可得出

$$np = n_i^2 \exp(qV/kT) \tag{3.40}$$

将式 (3.40) 代入式 (3.19) 得

$$U_{SRH} = n_i^2 [\exp(qV/kT) - 1] / \{n + p + 2n_i \cosh[(E_t - E_i)/kT]\tau_0\}$$
$$\approx n_i^2 \exp(qV/kT) / \{n + p + 2n_i \cosh[(E_t - E_i)/kT]\tau_0\} \tag{3.41}$$

由于式 (3.41) 的分母在 $n = p$ 且 $E_t = E_i$ 时达到最小值，因此最大 U_{SRH} 为

$$U_{SRH}^{max} \approx (n_i/2\tau_0) \exp(qV/2kT) \tag{3.42}$$

假设绝大多数耗尽层都具有 U_{SRH}^{max}，则最大 $J_{recombination}$ 可以近似为

$$J_{recombination}^{max} \approx qU_{SRH}^{max} W_D \tag{3.43a}$$
$$= J_S^R \exp(qV/2kT)$$

和

$$J_S^R = qW_D n_i / 2\tau_0 \tag{3.43b}$$

式 (3.43a) 表明，小注入条件下空间电荷复合电流的理想因子为2。

3.6.2 大注入条件

即使在大注入条件下，同样可以假定电荷中性条件在每个点都成立（即 $\partial\Delta p(x) \approx p(x)$ 等于 $\partial\Delta n(x) \approx n(x)$）。基于式 (3.32)，当 W_D 大于漂移层厚度 $2d$ 时，

$$p(2d) \approx n_i \exp(qV/2kT) \tag{3.44}$$

显然，J_{total} 与 exp($qV/2kT$) 大致成正比，表明大注入电流的理想因子为2，其比例常数由下式给出[12]：

$$J_S^{HL} = \{2qD_a N_C^{0.5} N_V^{0.5} \tanh(d/L_a)/L_a/[1 - 0.25\tanh^4(d/L_a)]^{0.5}\} \times$$
$$\exp[-(E_g + qV_M)/(2kT)] \tag{3.45}$$

式中，V_M 为漂移区上的电压降；D_a 为双极扩散系数，其定义为

$$D_a = 2D_n D_p / (D_n + D_p) \tag{3.46}$$

双极扩散长度 L_a 定义为

$$L_a = (D_a \tau_{HL})^{0.5} \quad (3.47)$$

式中，τ_{HL} 为大注入寿命。

3.6.3 测量电流/电压特性的示例

根据 3.6.1 节和 3.6.2 节所述的公式，J_{total} 可表示为

$$J_{total} = J_S^R \exp(qV/2kT) + J_S^{LL} \exp(qV/kT) + J_S^{HL} \exp(qV/2kT) \quad (3.48)$$

此处忽略串联电阻的影响，图 3.6 显示了一个直径为 60μm 的 GaN p⁺n 二极管电流/电压特性[13]，以及在式（3.48）中右侧的三个分量。二极管的层状结构由 p⁺GaN（$N_A = 2 \times 10^{19} cm^{-3}/0.5\mu m$），n⁻GaN（$N_D = 2 \times 10^{16} cm^{-3}/10\mu m$）和 n⁺GaN（$N_D = 2 \times 10^{18} cm^{-3}/2\mu m$）组成 GaN（0001）衬底。在此示例中，当电压 V 小于 2.8V 时，$n = 2$ 的空间电荷复合将主导正向电流。在 2.8~2.9V 的电压范

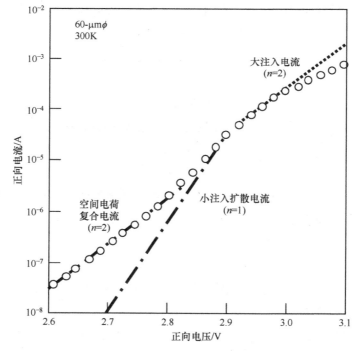

图 3.6 测得的 GaN p⁺n 二极管正向电流/电压特性示例[13]，
以及分段的电流分量：理想因子 n 为 2 的空间电荷复合，
$n = 1$ 的小注入和 $n = 2$ 的大注入

围内,电流分量由 $n=1$ 的小注入形成。而电压为 2.9~3.0V 时,正向电流由 $n=2$ 的大注入电流分量控制。

当 V 大于 3.0V 时,图 3.6 中的测量结果偏离了大注入电流分量曲线。通常将这种偏差称为串联电阻效应[14]。但是,在这个电压 V 范围内必须考虑多维效应(将在 3.7 节中进行描述)。

3.7 多维正向电流/电压特性

本节将阐明正偏 p^+n 结有关的多维效应。对于 Si p^+n 二极管而言,通常采用平面结构,通过离子注入和受主扩散形成 p^+ 区(见图 3.7a)。但是,平面结构不适用于 GaN 或 4H-SiC p^+n 二极管。因为虽然可以将镁注入 GaN[15-17]中,但镁受主的活化率非常低,这导致 p-n 结二极管的理想因子大(即 2.6),而且正向电流密度低(即 $1A/cm^2$)[17]。对于 4H-SiC 而言,经常使用的是铝注入,然而,离子注入尚未完全实现电导调制[18]。因此,在 GaN 和 4H-SiC p^+n 二极管中通常避免对有源区进行离子注入。

GaN 和 4H-SiC 的 p^+n 二极管的 p^+ 区通常由 n 型外延层上生长 p^+ 外延层而形成(见图 3.7b)。为了定义有源区的面积,需要形成台面结构(见图 3.7c)(第 6 章中将详细描述外延生长),第 7 章将详细介绍器件处理,包括刻蚀。制备 p^+n 二极管的工艺步骤分别为在 p^+n 二极管的背面和顶部形成阴极和阳极。在这种小型 p^+n 二极管中,会由于表面复合而导致外部电流增加。因为激活 GaN 中镁受主型掺杂需要退火,所以阳极电极通常采用非自对准工艺形成[19,20]。而电场集中也会引起非自对准台面型 p^+n 二极管外部电流的增加。

3.7.1 表面复合对 p^+n 二极管外部电流的影响

3.7.1.1 表面复合速度

3.4.2 节讨论了大部分半导体材料中均匀分布的 SRH 复合中心。类似地,在半导体表面上的带隙中也存在着局域态(E_{st})。因此,表面的面密度 [$N_{st}(cm^{-2})$] 将代替中心处的体积密度。当 $\sigma_n = \sigma_p \equiv \sigma$ 时,式(3.15)、式(3.16a)和式(3.16b)减小为

$$U_s = \sigma v_{th} N_{st}(n_s p_s - n_i^2)/(n_s + n_{1s} + p_s + p_{1s}) \tag{3.49a}$$

$$n_{1s} = n_i \exp[(E_{st} - E_i)/kT] \tag{3.49b}$$

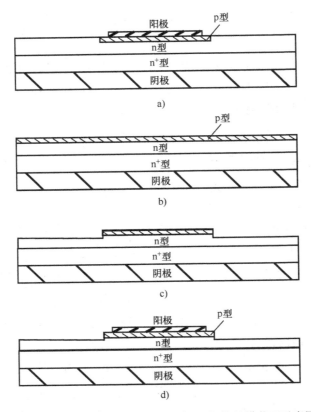

图 3.7 a) 平面型和 d) 台面型 p^+n 二极管的横截面示意图
以及 b) 和 c) 台面型 p^+n 二极管的制造工艺示意图

$$p_{1s}=n_i \exp[(E_i-E_{st})/kT] \quad (3.49c)$$

式中，n_s 和 p_s 分别为表面附近的电子和空穴浓度。

如果假设在 n 型半导体中整个表面空间电荷区域中的 np 乘积恒定：

$$n_s p_s = N_D^+ p' \quad (3.50)$$

式中，p' 为空间电荷区中性边缘处的空穴浓度。

当 $E_{st} \approx E_i$ 时，式（3.49a）减小为

$$U_s = \sigma v_{th} N_{st} N_D^+ [p'-(n_i^2/N_D^+)]/(n_s+p_s+2n_i) \quad (3.51a)$$
$$= s\Delta p'$$
$$s = \sigma v_{th} N_{st} N_D^+/(n_s+p_s+2n_i) \quad (3.51b)$$

式中，s 为表面复合速度；$\Delta p'$ 为空间电荷区中性边缘处的过剩空穴浓度。

如果表面处于电中性状态，则 $N_D^+/(n_s+p_s+2n_i)$ 为 1，因此 s 可以表示为

$$s = \sigma v_{th} N_{st} \quad (3.52)$$

3.7.1.2 表面复合影响模型

采用仿真工具[21]在圆柱坐标系中模拟了 3.6.3 节中描述的 GaN p$^+$n 二极管电流/电压特性（见图 3.8）。假设 n$^-$GaN 漂移层的过刻蚀深度为 $0.5\mu m$，并且使用的仿真参数与文献 [13] 和 [22] 中报道的相同。为了确定表面复合在测得的正向电流中起主导作用，将以下表面复合速度之一作为参数：s_1 在 p$^+$GaN 的侧面上；s_2 在 n$^-$GaN 的侧面上，同时 s_3 在刻蚀的 n$^-$GaN 表面上。

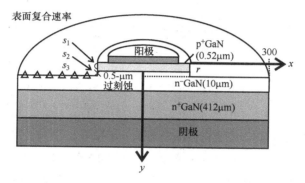

图 3.8 模拟的 GaN p$^+$n 二极管示意图，包括以下三个表面复合速度：s_1 在 p$^+$GaN 的侧面上；s_2 在 n$^-$GaN 的侧面上；同时 s_3 在刻蚀的 n$^-$GaN 表面上

如图 3.9 所示，在 $s_2 = 1\times 10^8 cm/s$ 的情况下，可以体现出测得的 2.7V 正向电流与台面直径的关系，这也证实了 s_2 模拟的电流单位斜率即为外部电流。此外，当 s_1 和 s_3 的值较大时（即 $1\times 10^8 cm/s$），采用 s_1 和 s_3 模拟也显示出较低的正向电流。图 3.9 中的斜率表明，小注入条件下，具有 s_1 和 s_3 的电流主要由体扩散电流控制。从而，GaN 台面型 p$^+$n 二极管的正向电流由 s_2 主导，即载流子复合发生在刻蚀 n$^-$GaN 的侧面上。

3.7.2 电场强度对非自对准台面型 p$^+$n 二极管的影响

接下来将分析大正向偏置 p$^+$n 结的多维效应。为了避免深受主在 GaN 和 4H-SiC 表面复合中的影响（请参见 2.4.2 节），在 300K 和刻蚀深度为零的条件下，对非自对准台面型 Si p$^+$n 二极管进行数值研究（见图 3.10）[23]。根据圆柱坐标系采用模拟工具模拟了二极管的电位分布（阳极-电极半径：$30\mu m$；台面

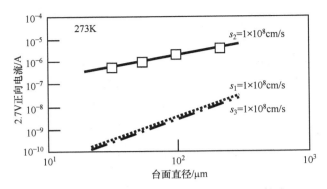

图 3.9　GaN p⁺n 二极管的实测值（空心正方形[13]）和
2.7V 正向电流随台面直径变化的曲线

半径：35μm)[21]。外延层由厚度为 0.5μm 的 Si（N_A：$5×10^{17}$cm^{-3}）和厚度为 10μm 的 Si（N_D：$2×10^{16}$cm^{-3}）组成。对 Si 衬底而言，假定厚度为 500μm，N_D 为 $5×10^{18}$cm^{-3}，电子和空穴具有相同的寿命（10μs）。

非自对准台面型 Si p⁺n 二极管的等势线如图 3.11 所示。当 V 为 0.6V 时，n⁻Si 层中的电势变化仅为 22mV（见图 3.11a）。在这种偏置条件下，n⁻Si 层处于准电中性状态，少数载流子扩散占主导地位。随着 V 的增加，电导调制变得更强，并且 n⁻Si 层中的电势增加［即 0.8V 时 27mV（见图 3.11b）和 0.9V 时 38mV（见图 3.11c）］。此时，n⁻Si 层不再是准电中性状态，漂移电流分量占 J_{total} 的主导地位。在 0.8V 时，n⁻Si 层中的电势落到 p-n 结上。

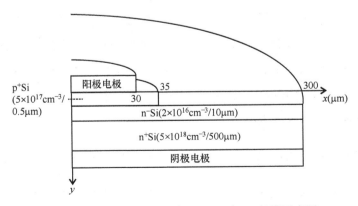

图 3.10　模拟的非自对准台面型 Si p⁺n 二极管示意图，
其中假设 n⁻Si 层的过刻蚀深度为零

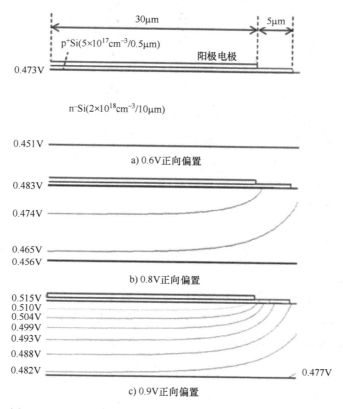

图 3.11 仿真得到的正向偏置电压分别为 a) 0.6V, b) 0.8V 和 c) 0.9V 时的非自对准台面型 Si p-n 结二极管的等势线

然而，当电压为 0.9V 时，n⁻Si 层中的电势延伸到 p⁺Si 层中，并且电势集中于阳极边缘的 p⁺Si 层中（见图 3.11c）。电场集中在阳极边缘导致了 J_{total} 峰值超过 $10^4 A/cm^2$（见图 3.12），因此 J_{total} 的显著增加成为非自对准台面型 p⁺n 结的特征之一，其在光子回收中起着重要作用（在第 4 章中将进行描述）。

图 3.12 正向偏压为 0.9V 时，仿真得到的非自对准台面型 Si p-n 结二极管的电流密度等值线

3.8 结击穿

当 p-n 结上的反向电压形成的电场大到一定程度时,会发生以下两种击穿。第一种击穿称为雪崩击穿,自由载流子从电场中获得能量,从而破坏晶体中的共价键(请参见 2.6 节)。第二种击穿称为齐纳击穿,电子在与其他粒子不发生相互作用的情况下由价带直接跃迁至导带(即带间隧穿)。器件发生齐纳击穿的击穿电压(BV)小于发生雪崩击穿时的击穿电压。换而言之,由式(3.30)和式(3.31)可知,足够小的 W_D(即足够大的 N_D)是发生带间隧穿的必要条件。

对于 Si p-n 结,可以通过击穿电压(BV)与温度(T)的关系来判断发生的击穿是雪崩击穿还是齐纳击穿。击穿电压随着温度的增加而增大,这是由于电子的平均自由程减小,因此发生雪崩击穿;然而,击穿电压随着温度的增加而减小,这是由于带间隧穿的价带电子流密度增大,因此发生齐纳击穿。但是,对于由 GaN 和 4H-SiC 制成的 p-n 结,由于 GaN 和 4H-SiC 材料的晶格缺陷密度相对较大,所以这种趋势未必适用。例如,对于 4H-SiC p-n 结,所报道的 BV 与 T 的关系随结的不同而不同[24]。

3.9 总结

本章将 GaN 和 4H-SiC p-n 结的载流子复合寿命和电流/电压特性与 Si p-n 结进行了比较。本章介绍的多维效应,首次在半导体教科书中介绍了非自对准台面型 p-n 结边缘的电场集中,此类电场集中会导致较大的局部电流密度,可以认为是在直接带隙半导体中发生光子回收的原因(请参见第 4 章)。

参 考 文 献

[1] Shockley, W., "The Theory of p-n Junctions in Semiconductors and p-n Junction Transistors," *Bell System Technical Journal*, Vol. 28, No. 3, 1949, pp. 435–489.

[2] Fick, A., "On Liquid Diffusion," *Journal of Membrane Science*, Vol. 100, 1995, pp. 33–38.

[3] Schenk, H. P. D., et al., "Band Gap Narrowing and Radiative Efficiency of Silicon-Doped GaN," *Journal of Applied Physics*, Vol. 103, 2008, pp. 103502-1–103502-5.

[4] Goldberg, Y., M. E. Levinshtein, and S. L. Rumyantsev, in *Properties of Advanced Semiconductor Materials GaN, AlN, SiC, BN, SiC, SiGe* (eds. Levinshtein, M. E., S. L. Rumyantsev, and M. S. Shur), New York: John Wiley & Sons, 2001, pp. 93–148.

[5] Trupke, T., et al., "Temperature Dependence of the Radiative Recombination Coefficient of Intrinsic Silicon," *Journal of Applied Physics*, Vol. 94, No. 8, 2003, pp. 4930–4937.

[6] Shockley, W., and W. T. Read, "Statistics of the Recombination of Holes and Electrons," *Physical Review*, Vol. 87, No. 5, 1952, pp. 835–842.

[7] Hall, R. N., "Electron-Hole Recombination in Germanium," *Physical Review*, Vol. 87, No. 2, 1952, pp. 387–388.

[8] Scheibenzuber, W. G., et al., "Recombination Coefficients of GaN-Based Laser Diodes," *Journal of Applied Physics*, Vol. 109, 2011, pp. 093106-1–093106-6.

[9] Galeckas, A., et al., "Auger Recombination in 4H-SiC: Unusual Temperature Behavior," *Applied Physics Letters*, Vol. 71, No. 22, 1997, pp. 3269–3271.

[10] Schroder, D. K, et al., *Semiconductor Material and Device Characterization* (Third Edition), New York: Wiley-IEEE, 2006, pp. 392.

[11] Ahrenkiel, R. K., et al., "Ultralong Minority-Carrier Lifetime Epitaxial GaAs by Photon Recycling," *Applied Physics Letters*, Vol. 55, No. 11, 1989, pp. 1088–1090.

[12] Baliga, B. J, *Fundamentals of Power Semiconductor Devices*, New York: Springer-Verlag, 2008, Chapter 5.

[13] Mochizuki, K., et al., "Influence of Surface Recombination on Forward Current–Voltage Characteristics of Mesa GaN p+n Diodes Formed on GaN Freestanding Substrates," *IEEE Transactions on Electron Devices*, Vol. 59, No. 4, 2012, pp. 1091–1098.

[14] Sze, S. M., and K. K. Ng, *Physics of Semiconductor Devices* (Third Edition), John New Jersey: Wiley & Sons, 2007, pp. 96–100.

[15] Feigelson, B. N., et al., "Multicycle Rapid Thermal Annealing Technique and Its Application for the Electrical Activation of Mg Implanted in GaN," *Journal of Crystal Growth*, Vol. 350, 2012, pp. 21–26.

[16] Anderson, T. J, et al., "Activation of Mg Implanted in GaN by Multicycle Rapid Thermal Annealing," *Electronics Letters*, Vol. 50, No. 3, 2014, pp. 197–198.

[17] Anderson, T. J, et al., "Improvements in the Annealing of Ion Implanted III-Nitride Materials and Related Materials," *International Conference on Compound Semiconductor Manufacturing Technology*, Miami, 2016, pp. 225–228.

[18] Kimoto, T., et al., "Promise and Challenges of High-Voltage SiC Bipolar Power Devices," *Energies*, Vol. 9, No. 11, 2016, pp. 908-1–908-15.

[19] Amano, H., et al., "Electron Beam Effects on Blue Luminescence of Zinc-Doped GaN," *Journal of Luminescence*, Vol. 40–41, 1988, pp. 121–122.

[20] Nakamura, S., T. Mukai, and M. Senoh, "High-Power GaN p-n Junction Blue-Light-Emitting Diodes," *Japanese Journal of Applied Physics*, Vol. 30, No. 12A, 1991, pp. L1998–L2001.

[21] http:// www.silvaco.com/products/device_simulation/atlas.html.

[22] Mochizuki, K., et al., "Numerical Analysis of Forward Current-Voltage Characteristics of Vertical GaN Schottky-Barrier Diodes and p-n Diodes on Freestanding Substrates," *IEEE Transactions on Electron Devices,* Vol. 58, No. 7, 2011, pp. 1979–1985.

[23] Mochizuki, K., "Vertical GaN Bipolar Device: Gaining Competitive Advantage from Photon Recycling," *Physica Status Solidi A,* Vol. 214, No. 3, 2017, pp. 160048-1–160048-8.

[24] Konstantinov, A. O., et al., "Temperature Dependence of Avalanche Breakdown for Epitaxial Diodes in 4H Silicon Carbide," *Applied Physics Letters,* Vol. 73, No. 13, 1998, pp. 1850–1852.

第 4 章

光子回收效应

4.1 引言

因为 GaN 是一种直接带隙半导体材料（参见 2.3 节），它的载流子复合寿命很短（参见 3.4 节），所以研究人员常认为 GaN 不适用于双极功率器件，正如 Baliga 在文献 [1] 中所述："对于 GaN 功率器件的关注仅限于单极功率器件"。不过 Baliga 在同一本书中也提到了文献 [2] 所报道的 3.7kV GaN p-n 结二极管，其描述为："当电流密度为 100A/cm² 时，这些二极管表现出了期望的 3.0V 膝点电压和 3.3V 导通压降。尽管在漂移区中测得的载流子寿命很短，仅为 2ns，但仍然能够获得较小的导通压降。"文献 [3] 虽然在书中没有说明导通压降较小的原因，但可以认为是由于光子回收而延长了载流子的复合寿命。为了改善 GaN p-n 结二极管（参见 10.3 节）和 GaN 双极晶体管的性能（参见 10.5.4 节），需要充分利用光子回收的优势。

如在 2.3 节和 3.4 节中所述，光子回收是对直接带隙半导体（例如 GaN、GaAs 和 $Al_xGa_{1-x}As$ （$x<0.4$）[4,5]）中复合辐射的重新吸收。因此，它在间接带隙半导体（例如 SiC 和 Si）中的影响可以忽略不计。对于 AlGaAs-GaAs 异质结构，文献 [6-8] 已经研究了光子回收对辐射复合寿命[6]、扩散长度和内量子效率[7]以及少数载流子寿命[8]的影响。1990 年和 1991 年，国际上分别在半导体器件中首次利用光子回收降低脊波导激光二极管[9]和表面发射激光二极管[10]的阈值电流。2011 年，光子回收被用于改善 GaAs 太阳能电池和垂直型 GaN p-n 结二极管的性能，实现了 28.2% 的转换效率（η），打破了之前 26.4% 的记录，并且还实现了与温度无关的较小的差分特征导通电阻 $R_{on}A^{diff}$[12]（见 1.3.2

节)。此后,光子回收被应用于多结太阳能电池与击穿电压大于 3kV 的 GaN p-n 结二极管。例如,双结非聚光太阳能电池的转换效率 η 与 GaN p-n 结二极管的优值 $BV^2/R_{on}A^{diff}$ 在 2013 年以后都迅速偏离各自原来的发展趋势[13-22](见图 4.1)。不过太阳能电池与 p-n 结二极管中的光子回收并不相同。本章将光子回收分为几种类型,并根据实验结果对它们进行建模。

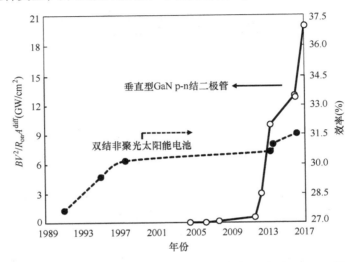

图 4.1 双结非聚光太阳能电池的转换效率与垂直型 GaN p-n 结二极管的优值(击穿电压(BV)的平方与微分电阻($R_{on}A^{diff}$)的比值)随年份变化曲线

4.2 光子回收现象的分类

光子回收的类型可以通过发射光子的回收方式来区分(见图 4.2)。对于外部光子回收,如图 4.3a 所示,从有源区 1 向外发射的部分光子(能量为 $h\nu_1$)在带宽较小的有源区 2 中被回收,即有源区 2 吸收了从有源区 1 发射的光子,从而产生新的电子空穴对,新产生的电子空穴对迅速发射声子(即能带边缘的热作用),同时,导带底(E_{C2})电子与价带顶(E_{V2})空穴的复合发射能量为 $h\nu_2$ 的光子(其中 $h\nu_2$ 小于 $h\nu_1$)(见图 4.3b)。

通过在有源区 1 中使用蓝光谱

图 4.2 光子回收现象的分类

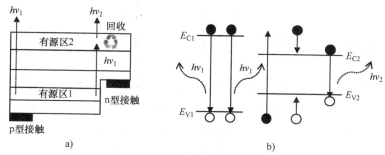

图 4.3 a) 半导体光子回收发光二极管截面图[23] 和
b) 用于说明外光子回收过程的能带图

InGaN 和在有源区 2 中使用黄光谱 AlGaInP，1999 年首次实现了外部光子回收的白光发光二极管（LED）[23]。关于外部光子回收的最新研究成果包括了安全的紫外线（UV）LED。由于紫外线是肉眼不可见的，并且会对人的皮肤和眼睛造成不可恢复的伤害，因此在操作过程中，将绿光发射作为指示和警告信号加入紫外线 LED 中。在有源区 1 中使用紫外光谱 $In_{0.01}Ga_{0.99}N$，并且在有源区 2 中使用绿色光谱 $In_{0.28}Ga_{0.72}N$，共同实现了具有外部光子回收绿色指示信号的紫外线 LED[24]。

内部光子回收指的是所发射的光子在主晶体内被回收，根据跃迁方式的不同，内部光子回收可以分为本征光子回收（IPR）和非本征光子回收（EPR），其中本征光子回收的跃迁方式为带间跃迁，非本征光子回收的跃迁方式为禁带中能级参与的跃迁[25]。如图 4.4 所示为一个正偏的直接带隙半导体 p^+-n 结，本征光子回收可以提高少子的寿命 τ_p，即将少数载流子有效寿命 τ_{eff} 与 τ_p 的比值定义为光子回收因子 Φ[6]（请参见 3.4.4 节）。请注意，对于特定的器件结构，Φ 应该具有不同的值。例如，对于 AlGaAs/GaAs/AlGaAs 的双异质结结构，其中 GaAs 层的厚度为 8μm，其 Φ 值为 4~6。另一方面，深受主或深施主能级的非本征光子回收增加了这些深能级杂质的电离率。但是，非本征光子回收的机理仍未建立，第 4.7 节中将介绍基于实验的结果得到的可能模型。

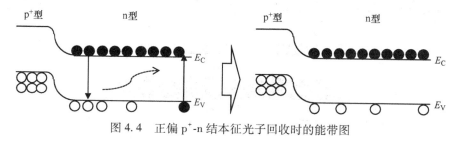

图 4.4 正偏 p^+-n 结本征光子回收时的能带图

4.3 本征光子回收

Velmre 与 Udal 评估了本征光子回收对垂直型 GaN p-n 结二极管的影响[26]。但是在讨论他们的研究工作之前，我们先研究 GaAs 太阳能电池的本征光子回收效应。通过将理想的扩散电流（理想因子 n 为 1）等效为光电流 I_{ph}，开路电压 V_{OC} 可以表示为[27]

$$V_{OC} = (kT/q)\ln(I_{ph}/I_S) \tag{4.1}$$

式中，k 为玻尔兹曼常数；T 为绝对温度；q 为电荷量；I_S 为饱和电流。

对于 p^+-n 型和 n^+-p 型的二极管，I_S 分别表示为[27]

$$I_S = (AqN_C N_V/N_D)(D_p/\tau_p^{eff})^{0.5}\exp(-E_g/kT) \tag{4.2a}$$

和

$$I_S = (AqN_C N_V/N_A)(D_n/\tau_n^{eff})^{0.5}\exp(-E_g/kT) \tag{4.2b}$$

式中，A 为有源区面积；N_C 为导带有效状态密度；N_V 为价带有效状态密度；N_D 为施主浓度；N_A 为受主浓度；D_p 为空穴的扩散率；D_n 为电子的扩散率；τ_p^{eff} 为空穴有效寿命；τ_n^{eff} 为电子有效寿命；E_g 为禁带宽度。

随着本征光子回收的影响增强，τ_p^{eff} 和 τ_n^{eff} 增大，从而降低 I_S 并且提高 V_{OC}。

例如，2011 年实现了转换效率优异的 GaAs 单结太阳能电池，它由前金属接触层、抗反射涂层、背金属接触层和柔性衬底组成（见图 4.5），其中，I_S/A 为 6.0×10^{-21} A/m²[28]。根据式（4.2a），在 $\Phi = 10$ 的简单假设下，当本征光子回收效应忽略不计时，I_S/A 应为 $6.0 \times 10^{-21} \times 10^{0.5} = 1.9 \times 10^{-20}$（A/m²）（见图 4.6a）。图 4.6b 显示了在光照下 GaAs 太阳能电池的电流/电压特性，其中实线和虚线分别表示有无本征光子回收时的特性。不过本征光子回收只对 V_{OC} 有影响，对于短路电流 I_{SC} 没有影响。因此，含有本征光子回收效应的最大输出功率大于不含本征光子回收效应的最大输出功率（在图 4.6b 中，前者为虚阴影线填充的长方形区域面积

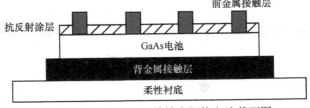

图 4.5 报道的 GaAs 单结太阳能电池截面图

$V_\mathrm{m}^\mathrm{w.o.PR} I_\mathrm{m}$,后者为实阴影线填充的长方形区域面积$(V_\mathrm{m} - V_\mathrm{m}^\mathrm{w.o.PR}) I_\mathrm{m}$。

对于2013年报道的双结太阳能电池（见图4.1），其转换效率 η 的提高与本征光子回收密切相关[29]。如图4.7所示，底部GaAs电池中辐射复合发射光子（如实心箭头所示），其能量约为GaAs的禁带宽度。由于在顶部电池（由带宽更大的InGaP组成）中不会吸收这些光子，因此能够实现高效的光子回收。但是，必须注意，InGaP电池中辐射复合发射的光子（虚线箭头所示）会在GaAs电池中被大量吸收，这种吸收降低了InGaP电池中本征光子回收的能力。

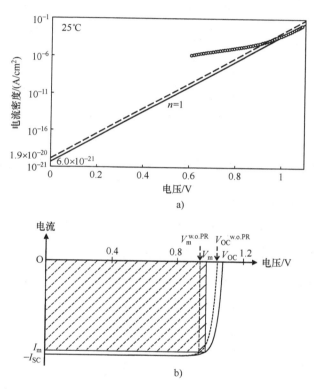

图4.6 a）暗电流密度/电压特性和 b）光照下的电流/电压特性。在图4.6a中，空心圆表示经串联电阻校正的数据[28]；实线和虚线分别表示含有和不含本征光子回收的理想扩散电流密度（理想因子 n 为1）。在图4.6b中，实线和虚线分别表示含有和不含本征光子回收的情况。I_m 和 V_m 是输出功率最大时对应的电流和电压值（I_SC：短路电流 [即光电流 I_ph]；V_OC：具有光子回收效应的开路电压；$V_\mathrm{OC}^\mathrm{w.o.PR}$：不具备光子回收效应的开路电压）

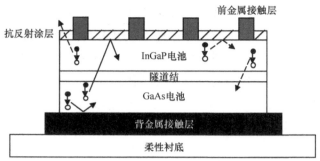

图 4.7　报道的 InGaP/GaAs 双结太阳能电池截面图[29]

Velmre 与 Udal 简单地分析了本征光子回收对垂直型 GaN p-n 结二极管的影响[26]。但是，他们的分析是在垂直型 GaN p-n 结二极管的正向电流（I_F）/电压（V_F）特性报道以前进行的。第 4.4 节将指出，仅考虑本征光子回收效应的仿真无法再现测得的垂直型 GaN p-n 结二极管的 I_F/V_F 特性。

4.4　本征光子回收对正偏 GaN p-n 结二极管的影响

在柱坐标系中，使用商用的器件仿真工具模拟了非自对准台面型 GaN p-n 结二极管的 I_F/V_F 特性[30]。台面结构如图 4.8 所示，其中阳极电极半径为 30μm，台面半径为 35μm。外延层由 0.5μm 厚的 GaN 层（受主浓度 N_A 为 5×10^{17}cm^{-3}；受主能级 E_A 为 0.235eV[31]）和 10μm 厚的 GaN 层（施主浓度 N_D 为 2×10^{16}cm^{-3}）组成。设 GaN 衬底的厚度为 500μm，浓度为 5×10^{18}cm^{-3}。使得电子和空穴的寿命 τ 相同，忽略阳极电极和阴极电极的接触电阻。

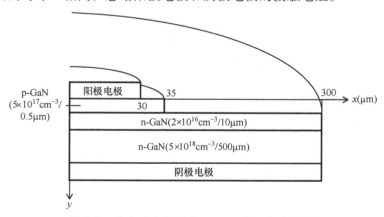

图 4.8　非自对准台面型 GaN p-n 结二极管结构

如图 4.9 所示，GaN p-n 结二极管的 I_F 随着 τ 的增加而增加。然而，当 τ 大于 0.5ms 时，I_F 饱和，此外，当 V_F 大于 3.6V 时，饱和的 I_F 小于测得的 I_F（见图 4.9 中的实心圆[18]）。这个结果表明，由于本征光子回收而延长的 τ 在确定垂直型 GaN p-n 结二极管的 I_F 时并不起主导作用。相比之下，即使 τ 为 50ns，仿真得到的 I_F 随着 E_A 的减小而单调增加。换句话说，当 E_A 在 0.135~0.185eV 之间时，仿真可以重现测得的 I_F（见图 4.10）。这个结果表明，正向电流的传导提高了 E_A 为 0.235eV 的深受主杂质的离化率。由于正向电流的传导不仅会产生复合辐射，还会产生热量，因此第 4.5 节将考虑自热效应对测得的 I_F 的影响。

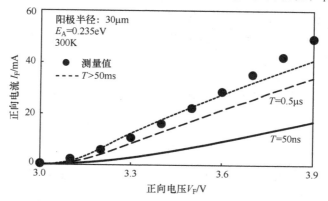

图 4.9 测得的 GaN p-n 结二极管正向电流/电压特性[18]
（实心圆），以及当少子寿命 τ 为 50ns（实线）、0.5μs（虚线）
和 50ms 或更长时间（点线）时仿真的正向电流/电压特性

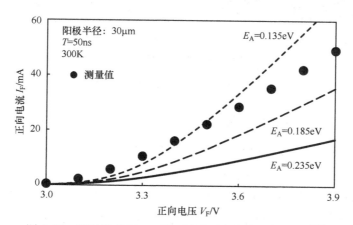

图 4.10 测量的 GaN p-n 结二极管正向电流/电压特性[18]
（实心圆），以及当受主能级 E_A 为 0.235eV（实线）、0.185eV
（虚线）和 0.135（点线）时仿真的正向电流/电压特性

4.5 自热效应对正偏 GaN p-n 结二极管的影响

Turin 和 Balandin 用图像法研究了 GaN 基大功率晶体管的自热效应[32]。他们计算了由点热源引起衬底（导热系数：κ_1；厚度：t）表面上的热扩散半径（b），该衬底的后表面与散热器（导热系数：κ_2）接触。当 $\kappa_2/\kappa_1 > 10$ 时，归一化热流的大小 $2b/t$ 与 κ_2/κ_1 无关，仅取决于净热流分数（见图 4.11 中的 Σ），即当 Σ 为 50%、60% 和 70% 时，$2b/t$ 分别为 2.0、2.3 和 2.8[32]。当热源包含一个有限的区域（半径：a）时，如图 4.11b 所示，只要热量均匀地在圆锥台上传播，就可以得到如下所示的沿垂直轴（z 轴）方向的圆形横截面（半径：b）上的热阻（R_{th}）[33]：

$$R_{th} = \int_{at/(b-a)}^{bt/(b-a)} (1-\kappa_1)t^2[\pi(b-a)^2]z^{-2}dz = [2/(\pi\kappa_1 a)]/(2b/t) \quad (4.3)$$

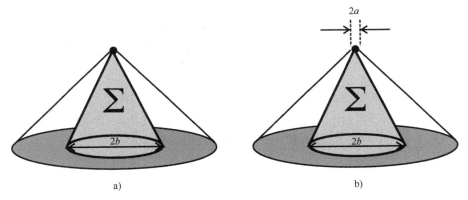

图 4.11　a) 点热源和 b) 圆形（半径：a）热源在半导体衬底表面产生的净热流分数 Σ 示意图

因此，如果 a 很小，可以估算出 GaN p-n 结二极管结温 T_j 的增量（ΔT_j）。当 a 为 30μm 时，将 $\kappa_1 = 2.0\text{Wcm}^{-1}\text{K}^{-1}$[34] 代入式（4.3）中得到：

$$R_{th}(\text{W/K}) = 106/(2b/t) \quad (4.4)$$

如图 4.9 和图 4.10 所示，在 300K 时，当二极管的正向偏压为 3.9V 时，二极管消耗的功率 P 为 0.192W。当从台面边缘向外辐射的光子比例为 η_e 时，增加 T_j 所需的净功率为 $(1-\eta_e)P$，因此 ΔT_j 可以表示为

$$\Delta T_j = 20\Sigma(1-\eta_e)/(2b/t) \quad (4.5)$$

由于确定 η_e 需要进行球积分测量,因此可以先计算出 ΔT_j 作为 η_e 的函数。如图 4.12 所示,在 $\eta_e=0$ 时,最大 ΔT_j 仅为 5.2K,并且几乎不受 Σ 变化的影响(Σ 的取值在 50%~70% 之间)。由于 ΔT_j 值太小,以至于在 300K 时无法增强受主杂质的电离。因此,第 4.4 节中所述的深受主杂质电离率的增加可以归因于辐射复合,即非本征光子回收。

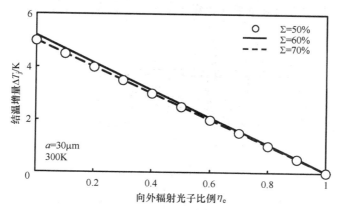

图 4.12　图 4.8 中出现的 GaN p-n 结二极管的
结温增量与向外辐射光子比例之间的关系
(净热流分数 Σ 在 50%~70% 之间变化)

4.6　非本征光子回收对正偏 GaN p-n 结二极管的影响

第 3.7 节指出,在正向偏压较大的非自对准 p-n 结二极管中,I_F 主要受外部电流的影响。因此,原本用于再现图 4.10 中 GaN p-n 结二极管 I_F 测量值的 E_A 只能用于确定有效受主能级 E_A^{eff}。为了定量确定由非本征光子回收产生的 E_A^{eff},Mochizuki 等人采用了具有矩形传输线(TLM)模式[35]的 GaN p^+-n 结外延层[36]。不同于文献[36]中使用的 $E_A=0.175\text{eV}$[37],最新报道的 E_A 为 0.235eV[31],以下内容将使用该 E_A 值。

TLM 模式的台面结构采用由 p^{++}-GaN(Mg:$2\times10^{20}\text{cm}^{-3}/20\text{nm}$)/$p^+$-GaN(Mg:$5\times10^{17}\text{cm}^{-3}/0.5\mu\text{m}$)/$n^-$-GaN(Si:$2\times10^{16}\text{cm}^{-3}/10\mu\text{m}$)/$n^+$-GaN(Si:$2\times10^{18}\text{cm}^{-3}/2\mu\text{m}$)组成的外延层结构,TLM 模式的台面结构是利用电感耦合等离子体(ICP)干法刻蚀制成的(见 7.2 节)。利用电子束沉积技术分别在 n^+-GaN

衬底的底面和 p^{++}-GaN 层的顶面形成钛/铝欧姆电极，以及钯/铝欧姆电极（见 7.6.1 节）。最后，以钯/铝电极为掩模板，用 ICP 干法刻蚀技术刻蚀掉 p^{++}-GaN 层裸露的部分。首先，利用传统方法测试 TLM 模式台面结构的电流/电压特性，即在 p-n 结两端加偏置电压，如图 4.13b 所示。假设间距最大（即 20μm）模式所对应的电阻主要由 p^+-GaN 层的方块电阻决定，则根据商用器件仿真工具[30]所得到的电流/电压特性，可以确定空穴迁移率 μ_p 为 15cm^2/Vs（见图 4.15）。

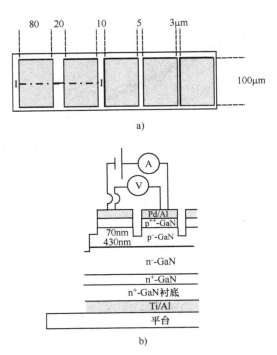

图 4.13 a) TLM 模式的平面示意图和 b) 传统的 TLM 测量装置沿直线 I-I 的截面图

然后其中一个钯/铝电极（阳极）相对于底部电极（阴极）施加一个正偏压，并测量两侧电极之间的电流/电压特性（见图 4.14）。如图 4.16 所示，在图 4.14 中的 3μm 间距/5μm 间距模式下，使用 90mA 阳极电流模拟仿真的电势为 V_1 和 V_2。当 $V_1-V_2>0.5$ 时，V_1 为常数，V_2 随 $|V_1-V_2|$ 的增加而线性减小，在上述条件下，可以得到 p^+-GaN 层中 5μm 间距区域的横向电流/电压特性。另一方面，在图 4.16 中，当 $V_1-V_2<-0.5$V 时，V_2 为常数，V_1 随 $|V_1-V_2|$ 的增加而线性减小，在上述条件下，可以得到 p^+-GaN 层中 3μm 间距区域的 $|I_L|/|V_1-V_2|$ 特性。

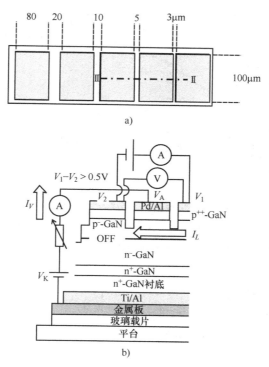

图 4.14　a）传输线（TLM）模式的平面示意图和
b）正向偏置的 TLM 测量装置沿直线 Ⅱ-Ⅱ 的截面图

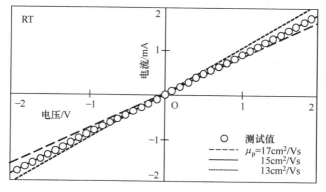

图 4.15　20μm 间距传输线模式器件中，电流/电压
特性的测试结果和仿真结果

当 $|V_1-V_2|<0.5\text{V}$ 时，V_1 和 V_2 都大于 p-n 结的内建电势 V_{bi}（约 3V），因此不仅在 p^+ 层中存在空穴电流，而且在 n^--GaN 层中也存在空穴电流。由于分析困难，所以未使用该条件下（$|V_1-V_2|<0.5\text{V}$）的实验结果。

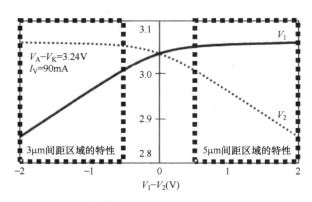

图4.16 3μm间距和5μm间距模式下,阳极电流为90mA时(即阳极与阴极的电势差(见图4.14中的V_A-V_K)为3.24V),仿真得到的电势V_1和V_2(见图4.14)随V_1-V_2的变化

纵向阳极电流I_V为90mA时,对于TLM模式的器件,测试和仿真得到的$|I_L|/|V_1-V_2|$特性如图4.17所示。当TLM模式中的间距为20μm时(见图4.17a),测得的I_L与使用0.235eV E_A模拟的I_L拟合得很好。然而,随着间距从10μm开始减小,测得的I_L与使用0.235eV E_A模拟的I_L之间的偏差增加,(见图4.17b~d)。由于这种趋势在I_V=10~70mA的情况下也是如此,因此可以得出结论:在10μm范围内,非本征光子回收效应从p型电极的边缘处开始迅速减退且与I_V无关。

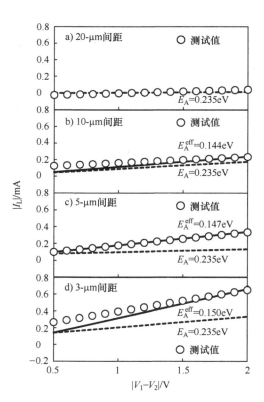

图4.17 纵向电流为90mA时,对于TLM模式器件,测试获得(带符号)和仿真获得的横向电流($|I_L|$)/横向电压($|V_1-V_2|$)特性

4.7 非本征光子回收的可能模型

对于电导调制 GaN p$^+$-n 结（请参见 3.7 节），光子主要通过 n$^-$-GaN 中的导带/价带跃迁产生。在文献 [33] 中，假设了光子的能量用于从电离的镁受主向导带发射电子（见图 4.18a-1），由此电中性受主通过俘获价带的电子而电离。电子从电离的镁受主发射到 p$^+$-GaN 导带相当于电子发射声子并移动到导带

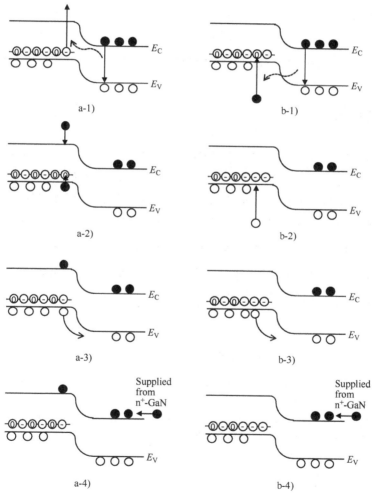

图 4.18 GaN p-n 结二极管中非本征光子回收的两种
可能模型（从 a-1~a-4 和从 b-1~b-4）的能带图

底部（见图 4.18a-2）。如果空穴注入 n^--GaN 中（见图 4.18a-3），同样数量的电子会从 n^+-GaN 注入 n^--GaN 中，以保持 n^--GaN 的电中性（见图 4.18a-4）。但是，如果电子与 p^+-GaN 中的空穴复合，则镁受主的离化比例不会增加。

中性镁受主利用光子的能量捕获了价带中的电子（见图 4.18b-1）实现受主电离，同时高能空穴在 p^+-GaN 导带中相当于空穴发射声子同时移至价带顶部（见图 4.18b-2）。在这种情况下，镁受主的离化比例增加。而且，如果将空穴注入 n^--GaN（见图 4.18b-3），为了保持 n^--GaN 的电中性，同样数量的电子也将从 n^+-GaN 注入 n^--GaN（见图 4.18b-4）。

如图 4.18a-1 ~a-4，图 4-18b-1 ~b-4 所示的两种情况均未违反动量守恒。由于受主和施主具有高度的局域性，因此它们的动量空间范围很广（根据不确定性原则）。但是，要揭示非本征光子回收的机理，还需要进一步的实验研究。

4.8 总结

本章将光子回收分为外部光子回收和内部光子回收，以白色和 UV 发光二极管为例介绍了外部光子回收的应用。内部光子回收近来被用于改善多结太阳能电池的 η 和垂直型 GaN p-n 结二极管的 $BV^2/R_{on}A^{diff}$，内部光子回收可进一步分为带到带跃迁的本征光子回收和涉及禁带能级跃迁的非本征光子回收。虽然本征光子回收可以改善 η，但单凭本征光子回收无法解释 $BV^2/R_{on}A^{diff}$ 的优化。因此，在 p^+-GaN 中，由深受主跃迁引起的非本征光子回收对提高深受主的电离率起着重要作用。然而，为了揭示非本征光子回收的机理，还需要进一步的实验研究。

参 考 文 献

[1] Baliga, B. J., *Gallium Nitride and Silicon Carbide Power Devices,* Singapore: World Scientific, 2017, p. 52.

[2] Kizilyalli, I. C., et al., "Vertical Power p-n Diodes Based on Bulk GaN," *IEEE Transactions on Electron Devices,* Vol. 62, No. 2, 2015, pp. 414–422.

[3] Baliga, B. J., *Gallium Nitride and Silicon Carbide Power Devices,* Singapore: World Scientific, 2017, p. 215.

[4] Dumke, W. P., "Spontaneous Radiative Recombination in Semiconductors," *Physical Review,* Vol. 105, No. 1, 1957, pp. 139–144.

[5] Stern, F., and J. M. Woodall, "Photon Recycling in Semiconductor Lasers," *Journal of Applied Physics,* Vol. 45, No. 9, 1974, pp. 3904–3906.

[6] Asbeck, P., "Self-Absorption Effects on the Radiative Lifetime in GaAs-GaAlAs Double Heterostructures," *Journal of Applied Physics*, Vol. 48, No. 2, 1977, pp. 820–822.

[7] Kuriyama, T., T. Kamiya, and H. Yanai, "Effect of Photon Recycling on Diffusion Length and Internal Quantum Efficiency in $Al_xGa_{1-x}As$-GaAs Heterostructures," *Japanese Journal of Applied Physics*, Vol. 16, No. 3, 1977, pp. 465–477.

[8] Ahrenkiel, R. K., et al., "Ultralong Minority-Carrier Lifetime Epitaxial GaAs by Photon Recycling," *Applied Physics Letters*, Vol. 55, No. 11, 1989, pp. 1088–1090.

[9] Gigase, Y. B., et al., "Threshold Reduction Through Photon Recycling in Semiconductor Lasers," *Applied Physics Letters*, Vol. 57, No. 13, 1990, pp. 1310–1312.

[10] Numai, T., et al., "Current Versus Light-Output Characteristics with No Definite Threshold in pnpn Vertical to Surface Transmission Electrophotonic Devices with a Vertical Cavity," *Japanese Journal of Applied Physics*, Vol. 30, No. 4A, 1991, pp. L602–L604.

[11] Savage, N., "Photon Recycling Breaks Solar Power Record," *IEEE Spectrum*, Vol. 48, No. 8, 2011, p. 16.

[12] Mochizuki, K., et al., "Photon-Recycling GaN p-n Diodes Demonstrating Temperature-Independent, Extremely Low On-Resistance," *International Electron Devices Meeting*, Washington, DC, Dec. 5–7, 2015, pp. 591–594.

[13] http://www.nrel.gov/ncpv/images/efficiency_chart.jpg.

[14] Irokawa, Y., et al., "Current–Voltage and Reverse Recovery Characteristics of Bulk GaN p-i-n Rectifier," *Applied Physics Letters*, Vol. 83, No. 11, 2003, pp. 2271–2273.

[15] Cao, X. A., et al., "Growth and Characterization of GaN PiN Rectifiers on Freestanding GaN," *Applied Physics Letters*, Vol. 87, 2005, pp. 053503-1–053503-3.

[16] Yoshizumi, Y., et al., "High-Breakdown-Voltage pn-Junction Diodes on GaN Substrates," *Journal of Crystal Growth*, Vol. 298, 2007, pp. 875–878.

[17] Nomoto, K., et al., "Over 1.0 kV GaN p-n Junction Diodes on Freestanding GaN Substrates," *Physica Status Solidi A*, Vol. 208, No. 7, 2011, pp. 1535–1537.

[18] Hatakeyama, Y, et al., "Over 3.0 GW/cm^2 Figure-of-Merit GaN p-n Junction Diodes on Freestanding GaN Substrates," *IEEE Electron Device Letters*, Vol. 32, No. 12, 2011, pp. 1674–1676.

[19] Hatakeyama, Y, et al., "High-Breakdown-Voltage and Low-Specific-on-Resistance GaN p-n Junction Diodes on Freestanding GaN Substrates Fabricated Through Low-Damage Field-Plate Process," *Japanese Journal of Applied Physics*, Vol. 52, 2013, pp. 028007-1–028007-3.

[20] Ohta, H., et al., "Vertical GaN p-n Junction Diodes with High Breakdown Voltage over 4 kV," *IEEE Electron Device Letters*, Vol. 36, No. 11, 2015, pp. 1180–1182.

[21] Nomoto, K., et al., "GaN-on-GaN p-n Power Diodes with 3.48 kV and 0.95 $m\Omega cm^2$: A Record High Figure-of-Merit of 12.8 GW/cm^2," *International Electron Devices Meeting*, Washington, D.C., Dec. 7–9, 2015, pp. 237–240.

[22] Ohta, H., et al., "5.0 kV Breakdown-Voltage Vertical GaN p-n Junction

Diodes," *Extended Abstracts of International Solid State Devices and Materials*, Sendai, Sep. 19–22, 20017, pp. 671–672.

[23] Guo, X., J. Grafi, and E. F. Schubert, "Photon Recycling Semiconductor Light Emitting Diode," *International Electron Devices Meeting*, Washington, D.C., Dec. 5–8, 1999, pp. 600–603.

[24] Chen, F. B., et al., "GaN-Based UV Light-Emitting Diodes with a Green Indicator Through Selective-Area Photon Recycling," *IEEE Transactions on Electron Devices*, Vol. 63, No. 3, 2016, pp. 1122–1127.

[25] Mochizuki, K., et al., "Influence of Surface Recombination on Forward Current–Voltage Characteristics of Mesa GaN p$^+$n Diodes Formed on GaN Freestanding Substrates," *IEEE Transactions on Electron Devices*, Vol. 59, No. 4, 2012, pp. 1091–1098.

[26] Velmre, E., and A. Udal, "Comparison of Photon Recycling Effect in GaAs and GaN Structures," *Estonian Academy of Sciences, Engineering*, Vol. 10, No. 3, 2004, pp. 157–172.

[27] Sze, S. M., and K. K. Ng, *Physics of Semiconductor Devices* (Third Edition), Hoboken, NJ: John Wiley & Sons, 2007, p. 723.

[28] Kayes, B. M., et al., "27.6% Conversion Efficiency, a New Record for Single-Junction Solar Cells Under 1 Sun Illumination," *37th IEEE Photovoltaic Specialists Conference (PVSC)*, Seattle, WA, June 19–27, 2011, pp. 4–8.

[29] Kayes, B. M., et al., "Flexible Thin-Film Tandem Solar Cells with >30% Efficiency," *Journal of Photovoltaics*, Vol. 4, No. 2, 2014, pp. 729–733.

[30] http://www.silvaco.com/products/tcad/device_simulation/atlas/atlas.html.

[31] Horita, M., et al., "Hall-Effect Measurements of Metalorganic Vapor-Phase Epitaxy-Grown p-Type Homoepitaxial GaN Layers with Various Mg Concentrations," *Japanese Journal of Applied Physics*, Vol. 55, 2016, pp. 05FH03-1–05FH03-4.

[32] Turin, V., and A. A. Balandin, "Electrothermal Simulation of the Self-heating Effects in GaN-Based Field-Effect Transistors," *Journal of Applied Physics*, Vol. 100, 2006, pp. 054501-1–054501-8.

[33] Mochizuki, K., et al., "Optical-Thermo-Transition Model of Reduction in On-Resistance of Small GaN p-n Diodes," *Japanese Journal of Applied Physics*, Vol. 52, 2013, pp. 08JN10-1–08JN10-4.

[34] Ohshima, Y., et al., "Thermal and Electrical Properties of High-Quality Freestanding GaN Wafers with High Carrier Concentration," *Physica Status Solidi C*, Vol. 4, No. 7, 2007, pp. 2215–2218.

[35] Berger, H. H., "Models for Contacts to Planar Devices," *Solid-State Electronics*, Vol. 15, No. 2-A, 1972, pp. 145–158.

[36] Mochizuki, K., et al., "Determination of Lateral Extension of Extrinsic Photon Recycling in p-GaN by Using Transmission-Line-Model Patterns Formed with GaN p–n Junction Epitaxial Layers," *Japanese Journal of Applied Physics*, Vol. 52, 2013, pp. 08JN22-1–08JN22-4.

[37] Pearton, S. J., C. R. Abernathy, and F. Ren, *Gallium Nitride Processing for Electronics, Sensors, and Spintronics*, London: Springer, 2006, p. 184.

第 5 章

体块单晶生长

5.1 引言

体块单晶生长对单晶晶圆的制造来讲至关重要。众所周知，硅晶圆是由多晶硅熔体制成。即使是 GaAs 材料，由于 As 具有较大的蒸气压，液态封装剂（例如氧化硼，化学式为 B_2O_3）辅助的熔体生长法已被广泛用于制造其单晶晶圆[1]。然而，并没有合适的封装剂可用于生长 GaN 或 SiC 体块单晶。

如图 5.1 中 GaN 和 SiC 的二元体系相图所示，在大气压下，GaN 或 SiC 不存在化学计量比的液相[2,3]。对于采用化学计量比熔体的液相法，理论上讲，

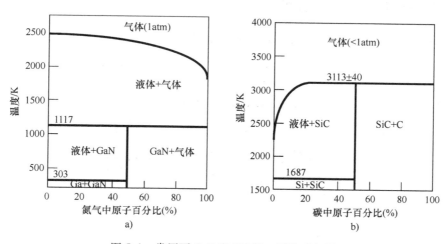

图 5.1 常压下 GaN 和 SiC 的二元体系相图

生长 GaN 单晶晶圆，其温度要大于 2700K，压力要大于 92 000atm⊖，如图 5.2 所示[2]；生长 SiC 单晶晶圆，温度要大于 3573K，压力要大约 98 000atm[3]。

因此，目前最常用的生长 GaN 和 SiC 体块单晶的技术均为气相生长法，即生长 GaN 单晶的氢化物气相外延（HVPE）法和生长 SiC 单晶的升华法。本章还将介绍生长体块单晶的其他技术，这些技术正在被深入地研究并用于提高晶体质量和生产效率。

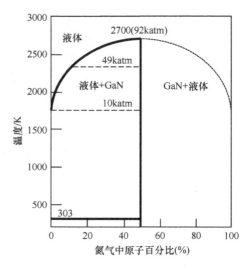

图 5.2 模拟计算的 GaN 凝聚态相图

5.2 HVPE 法生长 GaN

因为 GaN 籽晶极难获得（见 2.1 节），厚膜 GaN 通常利用 HVPE 法在异质衬底（如 GaAs 或蓝宝石）上生长。该生长方法最大的优势是具有较高的生长速率（通常是每小时几百微米）。

5.2.1 HVPE 法生长 GaN 的机制

氢化物气相外延反应腔通常包括源区和外延区两部分（见图 5.3）。

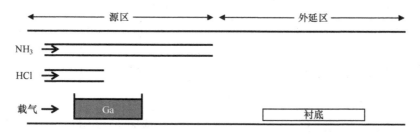

图 5.3 HVPE 法生长 GaN 的源区及外延区示意图

在源区，Ga 在特定温度下（大约 1100K）与 HCl 气体进行如下反应

$$2Ga(l) + 2HCl(g) \leftrightarrow 2GaCl(g) + H_2(g) \tag{5.1a}$$

⊖ 1atm=101.325kPa。——编辑注

$$Ga(l) + HCl(g) \leftrightarrow GaCl_2(g) + H_2(g) \quad (5.1b)$$

$$2Ga(l) + 6HCl(g) \leftrightarrow 2GaCl_3(g) + 3H_2(g) \quad (5.1c)$$

$$2GaCl_3(g) \leftrightarrow (GaCl_3)_2(g) \quad (5.1d)$$

公式括号中的"l"和"g"分别表示液相和气相。由上述反应在源区形成的气相化合物和 NH_3 分别由氢气、氮气、氦气、氩气等常用载气输运到外延区。

在外延区，上述的气相化合物将通过以下反应进行混合：

$$GaCl(g) + NH_3(g) \leftrightarrow GaN(s) + HCl(g) + H_2(g) \quad (5.2a)$$

$$2GaCl(g) + 2HCl(g) \leftrightarrow 2GaCl_2(g) + H_2(g) \quad (5.2b)$$

$$GaCl(g) + 2HCl(g) \leftrightarrow GaCl_3(g) + H_2(g) \quad (5.2c)$$

$$2GaCl_3(g) \leftrightarrow (GaCl_3)_2(g) \quad (5.2d)$$

公式括号中的"s"表示固相。根据已报道的热力学计算结果[4]，$GaCl_2$ 和 $(GaCl_3)_2$ 的气相分压非常小。此外，$GaCl_3$ 气相分压远低于 $GaCl$[5]，所以式 (5.2a) 被认为是最主要的反应。

GaN 在 HVPE 法生长情况下的过饱和度表示为

$$\sigma = (P^0_{GaCl} - P^e_{GaCl})/P^e_{GaCl} = (P^0_{GaCl}/P^e_{GaCl}) - 1 \quad (5.3)$$

式中，P^0_{GaCl} 和 P^e_{GaCl} 分别表示 GaCl 的输入与平衡分压。

5.2.2 GaN HVPE 法生长中的掺杂

二氯硅烷（SiH_2Cl_2）由于其高热稳定性，常用于硅掺杂 GaN 的 HVPE 法生长。值得注意的是，硅烷（SiH_4）在被输运到外延区之前就会分解为硅和氢，因此不适用于生长硅掺杂 GaN。另外在 HVPE 法中，金属镁已经被用于 p 型自支撑 GaN 的生长。

5.2.3 GaN 的横向外延生长

横向外延生长（ELO）技术可有效地弯曲从衬底延伸上来的位错，从而降低位错密度[8]。在 GaN 的生长中，传统的 ELO 可以通过两种方法进行改进：一是利用倒金字塔凹坑外延生长湮灭位错（DEEP）[9,10]，二是利用氮化钛（TiN）掩膜进行空洞辅助分离（VAS）[11,12]。

对于 DEEP[10] 来说，直接在 AsGa 表面形成具有圆形开口的二氧化硅（SiO_2）层。首先，在 AsGa 表面低温生长 GaN 缓冲层。随后在 NH_3 气氛下，将衬底温度升高至 GaN 的生长温度，生长出较厚的 GaN。当 GaN 层厚度超过

500μm 后，机械去除 GaAs 衬底。由于大量凹坑的存在，GaN 晶体中的位错将在每个凹坑的中心聚集，离凹坑中心越远的区域，位错密度越小。使用 150mm 的 GaAs 晶圆已经可以生长出同样直径的自支撑 GaN 衬底[13]。

相比之下，采用 VAS 法在蓝宝石衬底上制备的 GaN 衬底位错密度约为 $3\times10^6 cm^{-2}$ 且分布相对均匀。利用 VAS 法形成的 TiN 纳米掩膜，可以轻松地将 GaN 和蓝宝石衬底分离[11]。然而，由于晶圆的取向偏差，扩大晶圆直径的难度较大。最近，采用拼砌技术成功制备出 175mm 的自支撑 GaN 衬底，该技术将多层小晶圆与 HVPE 法[14]生长的厚 GaN 层合并在一起。

5.3 高压氮溶液生长 GaN

在高压氮溶液（HPNS）生长中，氮在液态镓中的溶解度相对较高，这是由于氮分子在镓表面分解并溶解在其中。在液态镓中，溶解的氮原子由高温区传输到低温区，在低温区 GaN 开始结晶生长。由于氮的溶解度非常低（小于 0.5%），GaN 的生长速率非常缓慢，因此晶体尺寸被限制在几毫米以内。然而，据报道 HPNS 生长的单晶具有非常低的位错密度（小于 $2\times10^2 cm^{-2}$）[15]。

5.4 钠助溶剂生长 GaN

通过向熔融镓中添加钠（Na），能够提高氮的溶解度。因此这种钠助溶剂法能够降低 GaN 的生长温度（800~1200K）和压力（小于 50atm）[16]。首先，氮分子在钠的作用下在气/液界面发生电离，被电离后的氮再溶解在 Ga-Na 熔体中。然后氮原子与镓原子结合，成核形成 GaN 晶体[16]。大多数氮原子在坩埚底部促进 GaN 籽晶的生长，但当氮原子浓度超过一个临界值时，就会发生液相外延（LPE）[17-23]。据报道，采用该方法[23]已经成功生长出低位错密度（小于 $10^3 cm^{-2}$）的 4in⊖ GaN 体块晶体[22]。

5.5 氨热法生长 GaN

氨热法生长 GaN 会在源区和 GaN 籽晶之间制造一个温度梯度。高压釜是高

⊖ 1in = 0.0254m。——编辑注

温下使用的高压容器,在高压釜中充满氨气,并加热到 800~900K,压力升高到 2000~4000atm,此时氨形成超临界流体(见图 5.4)。超临界液氨能将 GaN 从源区输送到籽晶区,继而生长[24]。为了提高超临界液氨中 GaN 的溶解度,通常使用碱性矿化剂(如 $NaNH_2$、$LiNH_2$ 和 KNH_2)或酸性矿化剂(如 NH_4Cl、NH_4Br 和 NH_4I)[25]。

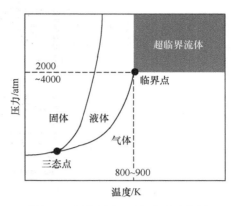

图 5.4 氨的压力-温度关系示意图

1995 年,Dwilinski 等人从镓、氨和 $LiNH_2$(或 KNH_2)中获得了高质量的 GaN 单晶[26]。此外,据报道用这种碱性矿化剂生长的 GaN 晶体的位错密度低于 $10^4 cm^{-2}$ [27]。另一方面,关于酸性矿化剂,使用酸性矿化剂生长 GaN 的试验结果在 2008 年首次报道,使用 HVPE 法生长的 GaN 籽晶,其位错密度较高($10^6 cm^{-2}$)[28]。

尽管氨热法生长 GaN 的生长速率在 c 面和 m 面分别提高到了 344μm/天和 46μm/天[29],但降低体块晶体中的杂质含量仍具有挑战性。例如,有文献报道,GaN 晶体中过渡金属和氧的浓度分别为低于 $10^{17} cm^{-3}$ 和 $10^{19} cm^{-3}$ 量级[5]。

5.6 升华法生长 SiC 单晶

1955 年,Lely 首次采用升华法生长出 SiC 单晶[30],该试验中很多片状 SiC 单晶在坩埚内随机自发成核。1978 年,Tairov 和 Tsvetkov 采用了籽晶升华法生长 SiC 单晶,该方法将籽晶固定在坩埚低温区[31]。这种改良后的 Lely 法将籽晶固定在坩埚顶部,SiC 粉料置于坩埚底部(见图 5.5),在氩气或氦气气氛中,使用射频感应或电阻加热的方式将坩埚加热到大约 2600K。

升华法生长的 4H-SiC 体块单晶直径不断增加,150mm 直径的晶圆已经实现商业化,200mm 直径的晶圆也有报道研制成功[32]。

5.6.1 SiC 的升华法生长原理

升华法生长体块 SiC 单晶的主要反应为[33]

$$Si_2C(g) + SiC_2(g) \leftrightarrow 3SiC(s) \tag{5.4a}$$

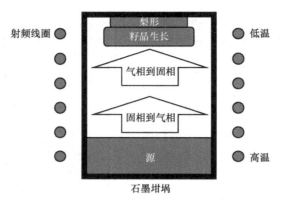

图 5.5　SiC 籽晶升华法坩埚示意图

$$\mathrm{SiC_2(g) + 3Si(g) \leftrightarrow 2Si_2C(g)} \tag{5.4b}$$

$$\mathrm{Si_2C(g) \leftrightarrow 2Si(g) + C(s)} \tag{5.4c}$$

$$\mathrm{Si(g) \leftrightarrow Si(l)} \tag{5.4d}$$

由于 SiC 堆垛层错形成能低（请参阅 2.2.3 节），因此生长过程中容易出现多型夹杂[34]。但是，通过优化工艺条件，即使是在重掺氮的条件下，也可以通过升华法生长没有多晶夹杂的 n 型 4H-SiC（000$\bar{1}$）体块单晶[35]。

5.6.2　升华法生长 SiC 单晶中的掺杂

SiC 的氮掺杂主要通过将氮气引入生长环境。SiC 晶体中的氮掺杂是由气相氮与生长表面吸附氮的平衡态决定的[36]。尽管我们可以把掺杂氮浓度增加到 $10^{20}\mathrm{cm^{-3}}$，但为了抑制堆垛层错的形成[37,38]，商用 n 型 4H-SiC 晶圆中的氮掺杂浓度通常控制在 $2\times10^{19}\mathrm{cm^{-3}}$ 以下。

另一方面，SiC 的 p 型掺杂可以通过在 SiC 粉料中添加铝来实现。铝在 SiC 晶体中的掺杂量几乎与坩埚中铝的蒸气压成正比关系。但是，高掺杂铝会促使 6H-SiC 晶型优先生长[35]，不利于晶型稳定，因此高铝掺杂目前仍然是一个挑战。

5.7　高温化学气相沉积法生长 SiC 单晶

为克服升华法生长 SiC 单晶的局限性[39-41]，业内开发了高温化学气相沉积（HT-CVD）法。在垂直坩埚中，SiC 体块单晶在温度为 2400~2700K，生长压力为 0.2~0.7atm 的工艺条件下进行生长（见图 5.6）。

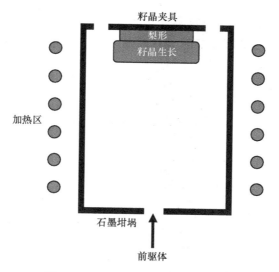

图 5.6　HT-CVD 法生长 SiC 示意图

高度过饱和蒸汽通过均匀成核的方式形成 Si 和 SiC 团簇。反应源气体（例如由 He 或 H_2 载气携带的 SiH_4 和 C_3H_8）通过加热区被引入反应腔顶部的籽晶托架上。典型的生长速率为 0.3~0.7mm/h。与升华法相比，HT-CVD 法的优势在于可以持续提供生长用的源材料，避免源材料耗尽，从而获得较长的 SiC 体块单晶。

5.8　溶液生长法生长 SiC 单晶

如 2.1 节所述，碳在硅溶液中的溶解度非常低，但在高压条件下其溶解度可以增大。例如，在 Ar 气氛 98atm 大气压条件下，可以利用溶液生长法生长 SiC 单晶[42]，但在温度 2500K 左右时的生长速率低于 0.5mm/h。

通过向 SiC 溶液中加入铬、钪或钛等，同样可以提高碳在硅溶液中的溶解度[43-49]。据报道，在温度为 2300K 时，SiC 的生长速率可达到 2mm/h。但是 SiC 晶体中的金属污染（大约为 $10^{17} cm^{-3}$ [50]）是一个很大的问题。在掺杂控制方面，在 Al-N 共掺杂条件下的 p 型和 n 型 4H-SiC 均可以成功生长[51]。

5.9　总结

图 5.7 比较了本章中提到的体块 GaN 晶体和 4H-SiC 单晶的生长速率、位错

密度和晶片的最大直径。采用拼砌技术将数量众多的钠助溶剂法生长的 GaN 薄片与 HVPE 法[14]生长的厚层 GaN 合并起来，可以制造大面积低位错密度的 GaN 晶圆。对于 4H-SiC 来讲，籽晶升华法是最成熟的生长方法，但是溶液法更适合生长高质量的 4H-SiC 单晶晶圆。

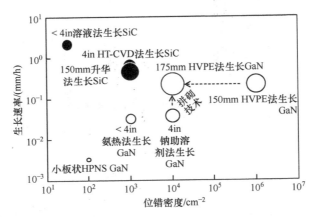

图 5.7 体块 GaN 晶体与 4H-SiC 单晶的生长速率、位错密度和晶片最大直径的对比（用圆圈表示直径的示意图代表了所研究晶圆的最大直径）

参 考 文 献

[1] Mullin, J. B, B. W. Straughan, and W. S. J. Brickell, "Liquid Encapsulation Techniques: The Use of an Inert Liquid in Suppressing Dissociation During the Melt-Growth of InAs and GaAs Crystals," *Journal of Physics and Chemistry of Solids*, Vol. 26, No. 4, 1965, 782–784.

[2] Unland, J., et al., "Thermodynamics and Phase Stability in the Ga–N System," *Bell System Technical Journal*, Vol. 28, No. 3, 1949, pp. 435–489.

[3] Tsvetkov, V. F., et al., "Recent Progress in SiC Crystal Growth," *Institute of Physics Conference Series*, Vol. 142, 1996, pp. 17–22.

[4] Koukitsu, A., et al, "Thermodynamic Analysis of Hydride Vapor Phase Epitaxy of GaN," *Japanese Journal of Applied Physics*, Vol. 37, No. 3A, 1998, pp. 762–765.

[5] Ke, X., W. Jian-Feng, and R. Guo-Qiang, "Progress in Bulk GaN Growth," *Chinese Physics B*, Vol. 24, No. 36, 2015, pp. 066105-1–066105-16.

[6] Richter, E., et al., "N-Type Doping of HVPE-Grown GaN Using Dichlorosilane," *Physica Status Solidi A*, Vol. 203, No. 7, 2006, pp. 1658–1662.

[7] Evanut, M. E., et al., "Incorporation of Mg into Thick Freestanding HVPE GaN," *MRS Advances*, Vol. 1, No. 2, 2016, pp. 169–174.

[8] Jastrzebski, L., J. F. Corboy, and R. Soydan, "Issues and Problems Involved

in Selective Epitaxial Growth of Silicon for SOI Fabrication," *Journal of the Electrochemical Society*, Vol. 136, No. 11, 1989, pp. 3506–3513.

[9] Motoki, K., et al., "Growth and Characterization of Freestanding GaN Substrates," *Journal of Crystal Growth*, Vol. 237–239, 2002, pp. 912–921.

[10] Motoki, K., "Development of Gallium Nitride Substrate," *SEI Technical Review*, No. 70, 2010, pp. 28–35.

[11] Ohshima, Y., et al., "Preparation of Freestanding GaN Wafers by Hydride Vapor Phase Epitaxy with Void-Assisted Separation," *Japanese Journal of Applied Physics*, Vol. 42, No. 1A/B, 2003, pp. L1–L3.

[12] Otoki, Y., et al., "Impact of Crystal-Quality Improvement of Eptaxial Wafers on RF and Power Switching Devices by Utilizing VAS-Method Grown GaN Substrates with Low-Density and Uniformly Distributed Dislocations," *CS MANTECH Conference*, May 2013, New Orleans, pp. 109–112.

[13] Kruszewski, P., et al., "AlGaN/GaN HEMT Structures on Ammono Bulk GaN Substrate," *Semiconductor Science and Technology*, Vol. 29, 2014, pp. 75004-1–75004-7.

[14] Yoshida, T., et al., "Development of GaN Substrate with a Large Diameter and Small Orientation Deviation," *Physica Status Solidi B*, Vol. 254, No. 8, 2017, pp. 1600671-1–1600671-4.

[15] Bockowski, M., "Bulk Growth of Gallium Nitride: Challenges and Difficulties," *Crystal Research and Technology*, Vol. 42, No. 12, 2007, pp. 1162–1175.

[16] Yamane, H., et al., "Preparation of GaN Single Crystals Using a Na Flux," *Chemistry of Materials*, Vol. 9, No. 2, 1997, pp. 413–416.

[17] Mori, Y., et al., "Growth of GaN Crystals by Na Flux LPE Method," *Physica Status Solidi A*, Vol. 207, No. 6, 2010, pp. 1283–1286.

[18] Kawamura, F., et al., "Growth of a Large GaN Single Crystal Using the Liquid Phase Epitaxy (LPE) Technique," *Japanese Journal of Applied Physics*, Vol. 42, No. 1A, 2003, pp. L4–L6.

[19] Kawamura, F., et al., "Novel Liquid Phase Epitaxy (LPE) Growth Method for Growing Large GaN Single Crystals: Introduction of the Flux Film Coated-Liquid Phase Epitaxy (FFC-LPE) Method," *Japanese Journal of Applied Physics*, Vol. 42, No. 8A, 2003, pp. L879–L881.

[20] Kawamura, F., et al., "The Effects of Na and Some Additives on Nitrogen Dissolution in the Ga-N System: A Growth Mechanism of GaN in the Na Flux Method," *Journal of Materials Science: Materials in Electronics*, Vol. 16, No. 1, 2005, pp. 29–34.

[21] Kawamura, F., et al., "Effect of Carbon Additive on Increases in the Growth Rate of 2 in GaN Single Crystals in the Na Flux Method," *Journal of Crystal Growth*, Vol. 310, No. 17, 2008, pp. 3946–3949.

[22] Mori, Y., et al., "Growth of GaN Crystals by Na Flux Method," *ECS Journal of Solid State Science and Technology*, Vol. 2, No. 8, 2013, pp. N3068–N3071.

[23] Imade, M., et al., "Growth of Bulk GaN Crystals By the Na-Flux Seed Technique," *Japanese Journal of Applied Physics*, Vol. 53, No. 5S1, 2014, pp. 05FA06-1–05FA06-5.

[24] Dwilinski, R., et al., "Properties of Truly Bulk GaN Monocrystals Grown by Ammonothermal Method," *Physica Status Solidi C*, Vol. 6, No. 12, 2009, pp. 2661–2664.

[25] Fukuda, T., and D. Ehrentraut, "Prospects for the Ammonothermal Growth of Large GaN Crystal," *Journal of Crystal Growth*, Vol. 305, No. 2, 2007, pp. 304–310.

[26] Dwilinski, R., et al., "GaN Synthesis by Ammonothermal Method," *Acta Physica Polonia A*, Vol. 88, No. 5, 1995, pp. 833–836.

[27] Dwilinski, R., et al., "Excellent Crystallinity of Truly Bulk Ammonothermal GaN," *Journal of Crystal Growth*, Vol. 310, No. 17, 2008, pp. 3911–3916.

[28] Ehrentraut, D., et al., "Reviewing Recent Developments in the Acid Ammonothermal Crystal Growth of Gallium Nitride," *Journal of Crystal Growth*, Vol. 310, No. 17, 2008, pp. 3902–3906.

[29] Pimpukar, S., et al., "Improved Growth Rates and Purity of Basic Ammonothermal GaN," *Journal of Crystal Growth*, Vol. 403, 2014, pp. 7–17.

[30] Lely, J. A., "Darstellung von Einkristallen von Silicium Carbid und Beherrschung von Art und Menge der eingebauten Verunreinigungen," *Berichte der Deutschen Keramischen Gesellschaft*, Vol. 32, 1955, pp. 229–236.

[31] Tairov, Y. M., and V. F. Tsvetkov, "Investigation of Growth Processes of Ingots of Silicon Carbide Single Crystals," *Journal of Crystal Growth*, Vol. 43, 1978, pp. 209–212.

[32] https://globenewswire.com/news-release/2015/07/16/752880/10142036/en/II-VI-Advanced-Materials-Demonstrates-World-s-First-200mm-Diameter-SiC-Wafer.html.

[33] Kaprov, S. Y., Y. N. Makarov, and M. S. Ramm, "Simulation of Sublimation Growth of SiC Single Crystals," *Physica Status Solidi B*, Vol. 202, No. 1, 1997, pp. 201–220.

[34] Knippenberg, W. F., "Growth Phenomena in Silicon Carbide," *Philips Research Reports*, Vol. 18, 1963, pp. 161–274.

[35] Kimoto, T., and J. A. Cooper, *Fundamentals of Silicon Carbide Technology*, Singapore: John Wiley & Sons, 2014, p. 49.

[36] Ohtani, N., et al., "Impurity Incorporation Kinetics During Modified-Lely Growth of SiC," *Journal of Applied Physics*, Vol. 83, No. 8, 1998, pp. 4487–4490.

[37] Chung, H. J., et al., "Stacking Fault Formation in Highly Doped 4H-SiC Epilayers During Annealing," *Materials Science Forum*, Vol. 433–436, 2003, pp. 253–256.

[38] Rost, H. J., et al., "Influence of Nitrogen Doping on the Properties of 4H-SiC Single Crystals Grown by Physical Vapor Transport," *Journal of Crystal Growth*, Vol. 257, No. 1–2, 2003, pp. 75–83.

[39] Kordina, O., et al., "High Temperature Chemical Vapor Deposition of SiC," *Applied Physics Letters*, Vol. 69, No. 10, 1996, pp. 1456–1458.

[40] Ellison, A., et al., "High Temperature CVD Growth of SiC," *Materials Science and Engineering B*, Vol. 61–62, 1999, pp. 113–120.

[41] Ellison, A., et al., "SiC Crystal Growth by HTCVD," *Materials Science Forum*,

Vol. 457–460, 2004, pp. 9–12.

[42] Hofmann, D. H., and M. H. Muller, "Prospects of the Use of Liquid Phase Techniques for the Growth of Bulk Silicon Carbide Crystals," *Materials Science and Engineering B*, Vol. 61–62, 1999, pp. 29–39.

[43] Syväyärvi, M., et al., "Liquid Phase Epitaxial Growth of SiC," *Journal of Crystal Growth*, Vol. 197, 1999, pp. 147–154.

[44] Kusunoki, K., et al., "Solution Growth of Self-Standing 6H-SiC Single Crystal Using Metal Solvent," *Materials Science Forum*, Vol. 457–460, 2004, pp. 123–126.

[45] Ujihara, T., et al., "Crystal Quality Evaluation of 6H-SiC Layers Grown by Liquid Phase Epitaxy Around Micropipes Using Micro-Raman Scattering Spectroscopy," *Materials Science Forum*, Vol. 457–460, 2004, pp. 633–636.

[46] Kamei, K., et al., "Solution Growth of Single Crystalline 6H, 4H-SiC Using Si–Ti–C Melt," *Journal of Crystal Growth*, Vol. 311, 2009, pp. 855–858.

[47] Danno, K., et al., "High-Speed Growth of High-Quality 4H-SiC Bulk by Solution Growth Using Si–Ti–C Melt," *Materials Science Forum*, Vol. 645–648, 2010, pp. 13–16.

[48] Daikoku, H., et al., "Top-Seeded Solution Growth of 4H-SiC Bulk Crystal Using Si-Cr Based Melt," *Materials Science Forum*, Vol. 717–720, 2012, pp. 61–64.

[49] Kado, M., et al., "High-Speed Growth of 4H-SiC Single Crystal Using Si-Cr Based Melt," *Materials Science Forum*, Vol. 740–742, 2013, pp. 73–76.

[50] Danno, K., et al., "Diffusion of Transition Metals in 4H-SiC and Trials of Impurity Gettering," *Applied Physics Express*, Vol. 5, No. 3, 2012, pp. 031301-1–031301-3.

[51] Mitani, T., et al., "4H-SiC Growth from Si–Cr–C Solution Under Al and N Co-doping Conditions," *Materials Science Forum*, Vol. 821–823, 2015, pp. 9–13.

第 6 章

外 延 生 长

6.1 引言

要想在 GaN 和 4H-SiC 功率器件中制造出设计所需的层状结构,外延生长是必不可少的(请参见 3.7 节)。本章将介绍 GaN 的金属有机化学气相沉积(MOCVD)和 4H-SiC 的化学气相沉积(CVD)的基本原理。此外,本章还将介绍 4H-SiC 的沟槽填充外延生长,这是用于高压 4H-SiC 超结(SJ)功率器件的一项先进技术(请参见第 8 章和第 9 章)。

6.2 GaN 金属有机化学气相沉积

MOCVD(也可称为金属有机物气相外延[MOVPE],有机金属化学气相沉积[OMCVD]或有机金属气相外延[OMVPE])是 III-V 族化合物半导体外延生长所用的最主要技术[1]。在 GaN 基材料中使用的 III 族前驱体与在 GaAs 和 InP 材料中使用的 III 族前驱体相同,即三甲基镓(TMG)、三乙基镓(TEG)、三甲基铝(TMA)和三甲基铟(TMI)。由于 V 族不同且前驱体的键能不同,GaN 基材料的 MOCVD 与 GaAs 基和 InP 基材料的 MOCVD 在生长温度上也有所不同。氨(NH_3)的键能(约 400kJ/mol)大于砷化氢(AsH_3)和磷化氢(PH_3)的键能(约 300kJ/mol)[2,3],因此,与 GaAs 基和 InP 基材料的生长温度(约 1200K)相比,GaN 基材料需要更高的生长温度(约 1400K)。

图 6.1 所示为用于 GaN 外延生长 MOCVD 系统的示意图。TEG 可以代替 TMG,作为金属有机源。在外延生长 AlGaN 和 InGaN 的时候,也可以使用 TMA

和 TMI 源。对于 n 型和 p 型掺杂,可以将 n 型和 p 型掺杂剂气体(例如,硅烷[SiH$_4$]和双环戊二烯基镁[CP$_2$Mg])的供应管线分别添加至 MOCVD 系统中。精确控制这些气体的流速,使得生长复杂器件的外延层成为可能。

GaN 的 MOCVD 生长通常在富含氮气的环境下进行。根据第一性原理计算和热力学分析,镓是通过气相镓原子掺入 GaN 生长表面的,这些镓原子的迁移是限速过程[4]。在实验中,GaN MOCVD 选区生长用于将具有螺旋位错的区域与没有螺旋位错的区域分开[5]。由螺旋位错引起的生长螺旋(见图 6.2)的级间距离 λ_0 的模型在文献[6]中给出。

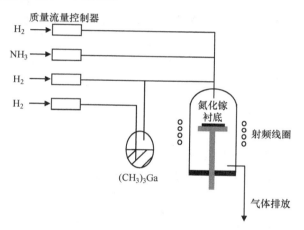

图 6.1 用于 GaN 外延生长 MOCVD 系统的示意图

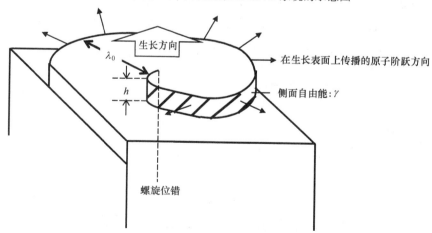

图 6.2 源自螺旋位错的生长螺旋的透视示意图。窄箭头表示在生长表面上传播的阶跃方向(h:螺旋台阶高度;γ:螺旋台阶侧面的自由能)

$$\lambda_0 = 19\gamma V_m / m\Delta\mu \tag{6.1}$$

式中，γ 为螺旋台阶侧面的自由能；V_m 为生长物的摩尔体积；m 为中心螺旋的数目（例如，图 6.3a 与 b 中的 $m=1$；图 6.3c 中的 $m=2$），$\Delta\mu$ 是生长气氛和晶体的化学势之差[7]。

$$\Delta\mu = RT\ln\alpha \quad (6.2)$$

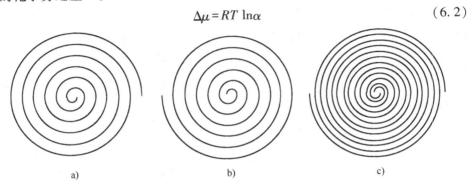

图 6.3 源自 a）与 b）一个和 c）两个螺旋错位的生长螺旋的平面示意图

式中，R 为理想气体常数；T 为生长温度；α 为过饱和比，定义为

$$\alpha = \sigma + 1 \quad (6.3)$$

式中，σ 为 5.2.1 节中定义的过饱和度。

Akasaka 等人绘制了在 GaN MOCVD 生长期间有螺旋位错和无螺旋位错区域的螺旋生长速率和晶核生长速率与 α 的函数关系图（由式（6.1）和式（6.2）计算得出），如图 6.4 中的空心圆符号所示[5]。通过基于 BCF 理论[8]计算得出的结果（图 6.4 中虚线）可以将他们的实验结果重现，该理论将在 6.4 节中进行描述。另一方面，在没有螺旋位错的情况下，生长速率非常低。该结果意味着，

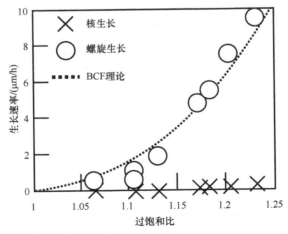

图 6.4 Akasaka 等人得到的螺旋生长速率与过饱和比的函数关系图[5]。
（空心圆符号和虚线分别代表实验生长速率和基于 BCF 理论计算得到的理论生长速率）[8]

α 低于二维成核的临界过饱和比 $α_{\text{critical}}$，这一点将在 6.3 节中进行说明。

6.3 二维成核理论

当 α 超过 $α_{\text{critical}}$ 时，成核作用占主导。本节将介绍基于二维成核理论的 $α_{\text{critical}}$ 的表达式。临界成核率表示为[9]

$$J_{\text{nuc}} = Z\omega_{\text{nuc}} n_{\text{nuc}} \tag{6.4}$$

式中，Z 为 Zeldovich 非平衡因子，用于解释由于临界核生长或分解而导致的临界核群数量耗尽；ω_{nuc} 为核生长达到超临界的频率；n_{nuc} 为临界成核浓度。式 (6.4) 可以重新排列为[9]

$$J_{\text{nuc}} = (2\pi r_{\text{nuc}} a v n_s n_0)(E_{\text{nuc}}/4\pi kT n_{\text{nuc}})^{0.5} \exp(-E_{\text{diff}}/kT)\exp(-E_{\text{nuc}}/kT) \tag{6.5}$$
$$\approx \exp(65)\exp(-E_{\text{nuc}}/kT)$$

式中，E_{nuc} 为晶核形成的自由能；n_{nuc} 和 r_{nuc} 分别为临界核的原子浓度和半径；a 为原子间距；v 为原子振动频率；n_s 为原子的表面浓度；n_0 为吸附原子位点的表面浓度；E_{diff} 为表面扩散的活化能。

对于圆盘状成核的情况，E_{nuc} 表示为[9]。

$$E_{\text{nuc}} = \pi h_1 \gamma^2 \Omega/(kT \ln α) \tag{6.6}$$

式中，h_1 为一个 Ga-N 或 Si-C 双层的高度（见图 6.5），$\pi h_1 \gamma^2$ 是原子的体积。结合式 (6.5) 和式 (6.6) 可得出著名的 $α_{\text{critical}}$ 公式：

$$α_{\text{critical}} = \exp\{\pi h_1 \gamma^2 \Omega/[(65-\ln J_{\text{nuc}})k^2 T^2]\} \tag{6.7}$$

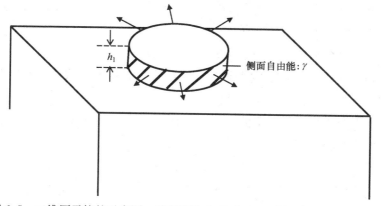

图 6.5　二维原子核的示意图，其高度为 GaN 或 SiC 双层，侧面自由能为 γ。箭头表示在生长表面上传播的台阶方向

6.4 BCF 理论

BCF 理论认为,生长若是在螺旋位错提供的台阶上进行,则表面扩散是一个限速过程[8]。这种理论已被应用于硅的气相外延[10,11]、富砷气氛下的 GaAs[12]、富镓气氛下的 GaAs[13]、GaAsSb 和 InGaAs[14]、富碳气氛下的 6H-SiC[15]和富硅气氛下的 4H-SiC[16]。本节将介绍基于 BCF 理论[8]的简单一维表面扩散模型(见图 6.6)。在原子向台阶扩散的过程中,一些原子到达台阶并融合到晶体中,而另一些则重新蒸发成蒸汽。

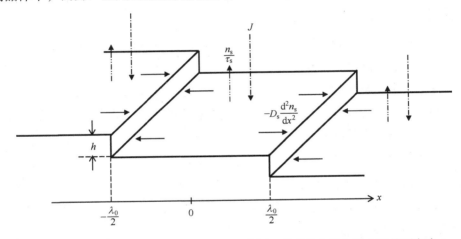

图 6.6 按照一维视图处理的螺旋台阶的示意图。高度为 h 的台阶之间的距离为 λ_0

当台阶上不发生成核时(即 $\alpha < \alpha_{critical}$,见 6.2 节),进入表面的净进入通量应等于朝向台阶的扩散通量,即

$$-D_s(d^2 n_s/dx^2) = J - n_s/\tau_s \tag{6.8}$$

式中,D_s 和 $n_s(x)$ 分别为吸附原子的表面扩散率和表面浓度;J 为进入的通量;τ_s 为吸附原子的平均停留时间。

如果假设台阶是吸附原子的完美吸收体(即原子在每个台阶的捕获概率都相同),则边界条件表示为

$$n_s(\pm \lambda_0/2) = n_{s0} \tag{6.9}$$

式中,n_{s0} 为平衡态时吸附原子的表面浓度。

那么,式(6.8)的解就变成了:

$$n_s(x) = J\tau_s + (n_{s0} - J\tau_s)[\cosh(x/\lambda_s)/\cosh(\lambda_0/2\lambda_s)] \tag{6.10}$$

式中，λ_s 为原子的表面扩散长度，详见文献 [8]。

$$\lambda_s(x) = (D_s\tau_s)^{0.5} = a\exp[(E_{des}-E_{diff})/kT] \quad (6.11)$$

式中，E_{des} 为脱附的活化能。生长速率 R_g 与 $(\sigma^2/\sigma_1)\tanh(\sigma_1/\sigma)^{[8]}$ 成正比。

$$\sigma_1 = 9.5\gamma a/(mkT\lambda_s) \quad (6.12)$$

即，当 $\sigma \ll \sigma_1$（见图 6.4）时，R_g 与 σ^2（即 $(\alpha-1)^2$）成正比，而当 $\sigma \gg \sigma_1$ 时，R_g 与 σ（即 $\alpha-1$）成正比。

当 $\lambda_s \gg \lambda_0$ 时，式（6.10）中的 $n_s(x)$ 变成抛物型（见图 6.7）。当吸附原子向台阶的通量（即 $-D_s(dn_s/dx)$）非常小，以至于 $x=0$ 处的最大 α 低于 $\alpha_{critical}$ 时，台阶上的二维成核不会发生。另一方面，当 $-D_s(dn_s/dx)$ 变得足够大以使最大 α 超过 $\alpha_{critical}$ 时，需要用二维成核理论来解释（见 6.3 节）。

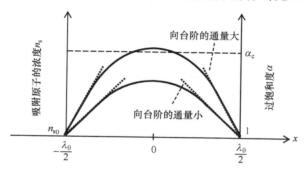

图 6.7　当 $\lambda_s \gg \lambda_0$ 时 $n_s(x)$ 和 $\alpha(x)$ 的分布

6.5　4H-SiC 的化学气相沉积

在 SiC 的 CVD 生长中，通常使用硅烷（SiH_4）和丙烷（C_3H_8）作为前驱体，氢气（H_2）作为载体。生长温度一般在 1800~1950K 之间。由于传统的水平[17]和垂直[18]构型的冷壁 CVD 反应器无法形成均匀的温度分布，因此，热壁 CVD 反应器被开发出来[19]。在使用热壁 CVD 反应器进行生长时（见图 6.8a~c），SiC 晶圆被放置在基座形成的气流通道内，射频感应可以对基座实现有效加热。由于晶圆是通过正面的热辐射和背面的热传导来加热的，因此，热壁 CVD 反应器很容易实现温度均匀性。

在 SiC CVD 生长的气相中，主要的化学反应如下[20]

$$SiH_4 \leftrightarrow SiH_2 + H_2 \quad (6.13a)$$

$$Si_2H_6 \leftrightarrow SiH_2 + SiH_4 \quad (6.13b)$$

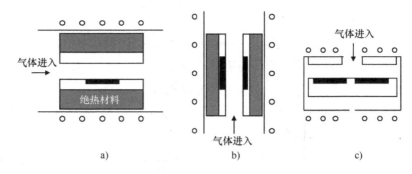

图 6.8 SiC 热壁 CVD 系统的示意图：a）水平式，b）烟囱式和 c）行星式反应器

$$SiH_2 \leftrightarrow Si+H_2 \qquad (6.13c)$$

$$2H+H_2 \leftrightarrow 2H_2 \qquad (6.13d)$$

$$C_3H_8 \leftrightarrow CH_3+C_2H_5 \qquad (6.13e)$$

$$CH_4+H \leftrightarrow CH_3+H_2 \qquad (6.13f)$$

$$C_2H_5+H \leftrightarrow 2CH_3 \qquad (6.13g)$$

$$2CH_3 \leftrightarrow C_2H_6 \qquad (6.13h)$$

$$C_2H_4+H \leftrightarrow C_2H_5 \qquad (6.13i)$$

$$C_2H_4 \leftrightarrow C_2H_2+H_2 \qquad (6.13j)$$

$$H_3SiCH_3 \leftrightarrow SiH_2+CH_4 \qquad (6.13k)$$

$$H_3SiCH_3 \leftrightarrow HSiCH_3+H_2 \qquad (6.13l)$$

$$Si_2 \leftrightarrow 2Si \qquad (6.13m)$$

$$Si_2+CH_4 \leftrightarrow Si_2C+2H_2 \qquad (6.13n)$$

$$SiH_2+Si \leftrightarrow Si_2+H_2 \qquad (6.13o)$$

$$CH_3+Si \leftrightarrow SiCH_2+H \qquad (6.13p)$$

$$SiCH_2+SiH_2 \leftrightarrow Si_2C+2H_2 \qquad (6.13q)$$

对于 n 型和 p 型掺杂，分别加入 n 型和 p 型掺杂气体（如氮气和 TMA）的供应管线。本节介绍了 4H-SiC CVD 的基本原理。第 6.6 节将介绍一种先进的沟槽填充 CVD 生长技术。

如 2.1 节所述，Kuroda 等人提出了一种用于重现与衬底相同的多型外延层的阶梯控制外延方法[21]。在该研究中，他们考察了 SiC（0001）衬底上各种取向的多型 SiC 层，并实现了 6H-SiC 的同质外延。此外，Kong 等人还报道了 6H-SiC 在取向杂乱的 6H-SiC 衬底上的同质生长[22,23]。对于 4H-SiC 而言，类似的同质外延生长方法也已被实现[24]。

据报道，台阶出现在取向错乱的衬底表面。由于这些台阶类似于源于螺旋位错的螺旋台阶（见 6.2 节），二维成核理论（见 6.3 节）和 BCF 理论（见 6.4 节）同样可以应用于处理取向错乱的衬底台阶的表面扩散问题[10-16]。当 $\lambda_s \ll \lambda_0$ 时，晶体生长最初是通过二维成核在阶梯上产生的（见图 6.9a）。如 2.2.2 节所述，主要由生长温度决定的多型晶体，在该条件下将变为 3C 型。3C-SiC 的生长有两种堆叠顺序，即 ABCABC……和 ACBACB……。另一方面，当 $\lambda_s \gg \lambda_0$ 时，吸附原子迁移并到达台阶（见图 6.9b）。取向错乱的 4H-SiC (0001) 的主导台阶有两倍或四倍的 SiC 双层高度[25]，这导致台阶处的加入位点 A、B、C 被唯一确定，由此实现外延层中衬底多型晶体的复制。

根据对 SiH_4-C_3H_8-H_2 生长体系中表面质量通量的模拟[17,26]，硅吸附原子与乙炔（C_2H_2）的表面反应为

$$2Si+C_2H_2 \leftrightarrow 2SiC+H_2 \qquad (6.14)$$

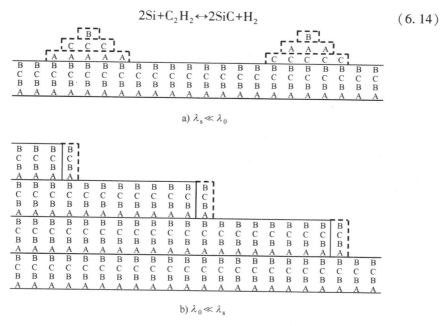

a) $\lambda_s \ll \lambda_0$

b) $\lambda_0 \ll \lambda_s$

图 6.9 当 a) $\lambda_s \ll \lambda_0$ 和 b) $\lambda_s \gg \lambda_0$ 时，在阶梯和四层台阶的情况下，表面生长的外延层的六方密堆结构中的占据位点（A、B 和 C）示意图

这是其中最活跃的表面反应之一。基于式（6.14），Kimoto 和 Matsunami[15] 利用二维成核（见 6.3 节）和 BCF（见 6.4 节）理论分析了硅吸附原子在富碳气氛下的表面扩散。然而，在该分析中，(0001) 表面的自由能被错误地应用于二维核的侧表面。因此，下文将分析富硅气氛下 C_2H_2 分子的表面扩散[16]。

Saito 和 Kimoto[27]在富硅气氛（即 C/Si 比［碳硅原子比］为 0.5~0.75）下，在 T = 1773K、生长压力为 80 Torr⊖的 SiH$_4$（1.5~2.0sccm）-C$_3$H$_8$-H$_2$（8.0slm）-氩气（0.8slm）体系中，生长出了 4H-SiC（0001）衬底上的 4H-SiC，偏角 θ 为 1°~45°。其中，sccm 表示每分钟标准立方厘米，slm 表示每分钟标准升。在台阶-气流生长模式下，阶梯上不会发生成核（见 6.4 节），C$_2$H$_2$ 分子到表面的净通量等于流向台阶的扩散通量。由连续性方程式（6.8）的解，即式（6.10）可以得到 C$_2$H$_2$ 在 x 方向的通量 $J_s(x)$ 为

$$J_s(x) = -D_s(\mathrm{d}n_s/\mathrm{d}x) = \lambda_s [J-(n_{s0}/\tau_s)][\sinh(x/\lambda_s)/\cosh(\lambda_0/2\lambda_s)] \tag{6.15}$$

假设 C$_2$H$_2$ 分子从台阶的左右两侧扩散，由此计算台阶速度。由于一个 C$_2$H$_2$ 分子在台阶边缘提供了两个碳原子[14]，因此，R_g 由台阶速度和 $\tan\theta$（即 h/λ_0）的乘积得出[15]

$$R_g = (4h\lambda_s/n_0\lambda_0)[J-(n_{s0}/\tau_s)]\tanh(\lambda_0/2\lambda_s) \tag{6.16}$$

当 $\lambda_s \gg \lambda_0/2$ 时，R_g 变得与 λ_0 无关 $\{$即 $(2h/n_0)[J-(n_{s0}/\tau_s)]\}$。根据文献[27]，当 θ = 4°~45°时，R_g 对 θ 的依赖性很小，但当 θ = 1°时，R_g 会减小。因此，λ_s 小于或等于 $h/(2\tan 1°)$（即对于两层台阶高度为 14.4nm，对于四层台阶高度为 28.9nm）。如果 n_0 等于表面上硅吸附原子的位点浓度（1.21×10^{15}cm^{-3}），则 $J-(n_{s0}/\tau_s)$ 可由式（6.16）计算得出。图 6.10 显示了在 λ_s 为 12~36nm 时，$J-(n_{s0}/\tau_s)$ 对由式（6.16）计算出的 C/Si 比值的依赖性。当 λ_s = 12~18nm（两层台阶高度）和 λ_s = 24~36nm（四层台阶高度）时，$J-(n_{s0}/\tau_s)$ 的最小二乘系数与 θ = 1°~45°的最小二乘系数一致。这个 λ_s 的范围与上述估计值一致（即 14.4nm 和 28.9nm）。由图 6.10a 和 b 中的垂直轴截距绝对值得出了 C$_2$H$_2$ 分子的平衡脱附通量，n_{s0}/τ_s 为（0.5~1.2）×10^{14}cm^{-2}s^{-1}。该值对应于 C$_2$H$_2$ 分子的平衡蒸汽压（$P_{C_2H_2}^e$）为（0.4~1.0）×10^{-4}Pa，由 Knudsen 方程[14]可以计算得到，即

$$n_{s0}/\tau_s = P_{C_2H_2}^e/(2\pi m_{C_2H_2}kT)^{0.5} \tag{6.17}$$

式中，$m_{C_2H_2}$ 为单个 C$_2$H$_2$ 分子的质量。

在富含硅的大气中，成核发生在阶梯上，对于这种二维成核模型（见 6.3 节），硅原子的平衡蒸汽压必须与硅的输入压力相近，即硅的过饱和比（α_{Si}）

⊖ 1Torr = 133.322Pa。——编辑注

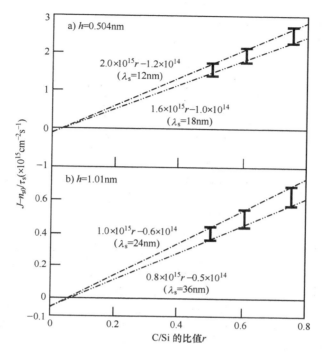

图 6.10 λ_s 为 12~36nm 时，$J-(n_{s0}/\tau_s)$ 对 C/Si 比值（由式（6.15）计算得出）的关系。用误差条表示的实验结果来自文献 [27]。直线和相应的方程代表了与实验结果相对应的最小二乘拟合结果

等于 $1^{[12]}$。由于碳（α_C）的过饱和比在台阶中心达到最大值（即 $x=0$）（见图 6.7），因此由式（6.10）可得 SiC 的最大过饱和比为

$$\alpha_{\max}=\alpha_{Si}\alpha_C(x=0)\approx 1+(\lambda_0 n_0 R_g/4h\lambda_s)(\tau_s/n_{s0})\tanh(\lambda_0/4\lambda_s) \quad (6.18)$$

当 $\alpha_{\max}$ 超过 α_{critical} 时，台阶上的成核模式就成为主导（见 6.3 节）。对于 $(10\times 10)\ \text{nm}^2$ 面积上每秒的圆盘状成核，式（6.7）变成了

$$\alpha_{\text{critical}}=\exp\{\pi h_1 \gamma^2 \Omega/[(65-\ln 10^{12})k^2 T^2]\} \quad (6.19)$$

式中，h_1 为一个 Si-C 层的高度（0.252nm）；Ω 为 Si-C 对的体积（$2.07\times 10^{-23}\text{cm}^3$）。

假定在富硅气氛下，二维原子核侧表面的自由能为 2.22J/m^2（这是根据硅终止 3C-SiC（111）计算出的 $\gamma^{[28]}$）。由式（6.18）和式（6.19）可计算出在 $T=1773\text{K}$ 时，在台阶-气流生长和二维成核两种模式之间转换的临界生长速率（R_c）对于偏角的依赖性（见图 6.11）。可以确定，R_c 的变化范围来源于 λ_s 的变化（见图 6.10），而不依赖于 h（$h=0.504$ 或 1.01nm）。还可以确定，文献

[27] 中的台阶-气流生长的实验数据在图 6.11 曲线的右下区域中，该曲线使用从图 6.10 获得的 n_{s0}/τ_s 和 λ_s 值计算得出。

根据式（6.18），R_c 随 n_{s0}/τ_s 的增加而增加，根据式（6.14）和式（6.17），n_{s0}/τ_s 的增加，即 $P^e_{C_2H_2}$ 的增加，可以通过降低 SiH_4 的流速来实现。然而，为了使 C/Si 比值保持在小于 1 的水平（即富含硅的气氛），必须限制 C_3H_8 的流动速率，从而降低 R_c。因此，图 6.11 中的计算曲线被认为是接近于 $T = 1773K$ 时的实际极限。

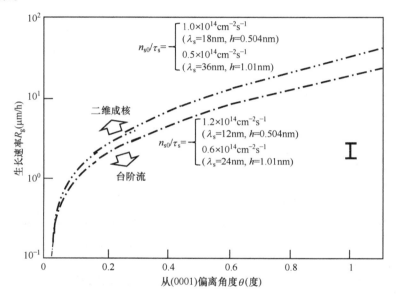

图 6.11 利用图 6.10 得到的 C_2H_2 分子的平衡脱附通量和表面扩散长度计算出的在台阶-气流生长（右下角区域）和二维成核（左上角区域）两种模式之间转换的临界生长速率对于偏角的依赖性。用误差棒表示的数据来自文献 [27]

6.6 4H-SiC 的化学气相沉积沟槽填充

如图 2.18 所示，具有 n 型漂移层的 p^+n 结的击穿电压（BV）等于 $E_{critical} W_D/2$（见图 6.12a 和图 6.12c 中的虚线）。如果在漂移层内有 n 型和 p 型区域交替出现（也就是所谓的 SJ 二极管）[29]，由于式（2.2）中的 N_D 实际上变成了零，所以同样的 BV 可以用厚度为一半（即 $W_D/2$）的漂移层来实现（见图 6.12b 和图 6.12c 中的实线）。因此，SJ 结构改善了 BV 与理想漂移层特征电阻

R_{ideal} 之间的权衡关系（见 2.7 节）。第一个 $BV=1.5$kV 的 4H-SiC SJ 二极管是通过多步外延法制造的[30]，即在图形化离子注入的 p 型区域上生长多个 n 型外延层，以形成交替的 p/n 柱。在硅的 SJ 结构中，由于外延生长过程中 p 型掺杂物（即硼）的受控扩散，p 型区域会发生合并[31]。然而，由于铝的扩散率非常小，硼的扩散率非常大，这种受控扩散在 SiC SJ 结构中是不可行的[30]。因此，对于 $BV>2$kV 的 4H-SiC SJ 功率器件来说，将外延层重新填充到 4H-SiC 沟槽中是更为实际的做法[32,33]。

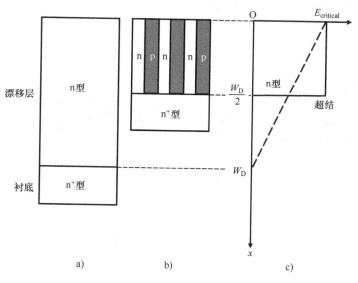

图 6.12　由 a) n 型半导体和 b) 超结结构组成的漂移层，以及 c) 它们的电场分布

在传统的 $SiH_4:C_3H_8:H_2$ CVD 系统中[33-37]，使用相对较高的压力（高达 38kPa[38]）可以减少与微沟槽尺寸相当的生长物质的平均自由程，从而可以使用连续流体近似[39]。在裸晶圆上生长速率较大的情况下（R_0），靠近非平面的生长物质的浓度等值线与表面平行，如图 6.13a 中的虚线所示。流向台面顶部的通量变大，而流向沟槽底部的通量变小，这导致表面形貌的不稳定。当 R_0 较小时（例如，由于生长物质在相对较高的温度 [1923K][33] 下的再蒸发），且曲率半径（r）较小时，由于二维吉布斯-汤姆逊效应[40,41]，浓度等值线的形状会发生变化，即

$$C^e(r) = C^e(\infty) \exp[\gamma V_m / RTr] \quad (6.20)$$

式中，$C^e(r)$ 为生长物质与曲率半径为 r 的曲面接触时的平衡气相浓度（见图 6.13b）。需要注意的是，r 的符号取决于台面是向上凸（$r>0$）还是向下凸

($r<0$)。

此外,在三维吉布斯-汤姆逊效应下[42],式(6.20)变为

$$C^e(r) = C^e(\infty)\exp[2\gamma V_m/RTr] \qquad (6.21)$$

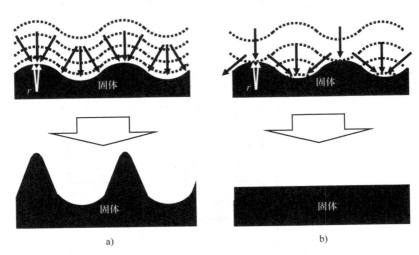

图 6.13 当生长速率和曲率半径(r)分别为 a)大和 b)小时,非平面生长表面的横截面示意图。虚线和实心箭头方向分别显示了蒸汽中生长物质的等浓度线和通量方向

为了进一步降低 R_0,氯化氢(HCl)气体被添加到常规 CVD 的生长步骤中[43-45]。这种情况下,流向台面顶部的通量变小,而流向沟槽底部的通量变大,从而形成均匀的表面。对于浅沟槽(小于 5μm 深)的情况,实验测得的生长速率(在台面顶部和沟槽底部)对沟槽间距的依赖性可以用式(6.20)计算得到[41,46]。在实验中,用 p 型 4H-SiC 来填充 25μm 深的 n 型 4H-SiC 沟槽,以此克服沟槽方向与[1120]方向轻微错位造成的强烈影响[47]。

6.7 总结

本章以二维成核和 BCF 理论为基础,阐述了 GaN MOCVD 生长和 4H-SiC CVD 生长的基本原理。同时,本章介绍了作为制造超结(SJ)结构的 4H-SiC 的沟槽填充外延生长的先进技术。然而,要将这种沟槽填充技术应用于 4H-SiC SJ 功率器件,发展评估 p 型掺杂物分布的技术是必不可少的。

参 考 文 献

[1] Dupuis, R. D., "III-V Semiconductor Heterojunction Devices Grown by Metal-organic Chemical Vapor Deposition," *IEEE Journal of Selected Topics in Quantum Electronics*, Vol. 6, No. 6, 2000, pp. 1040–1050.

[2] Simka, H., et al., "Computational Chemistry Predictions of Reaction Processes in Organometallic Vapor Phase Epitaxy," *Progress in Crystal Growth and Characterization of Materials*, Vol. 35, No. 2–4, 1997, pp. 117–149.

[3] Davidson, D. F., et al., "A Pyrosis Mechanism for Ammonia," *International Journal of Chemical Kinetics*, Vol. 22, No. 5, 1997, pp. 513–535.

[4] Sekiguchi, K., et al., "Thermodynamic Considerations of the Vapor Phase Reactions in III-Nitride Metal Organic Vapor Phase Epitaxy," *Japanese Journal of Applied Physics*, Vol. 56, 2017, pp. 04CJ04-1–04CJ04-4.

[5] Akasaka, T., Y. Kobayashi, and M. Kasu, "Step-Free Hexagons Grown by Selective-Area Metalorganic Vapor Phase Epitaxy," *Applied Physics Express*, Vol. 2, No. 9, 2009, pp. 091002-1–091002-3.

[6] Cabrera, N., and M. M. Levine, "On the Dislocation Theory of Evaporation of Crystals," *Philosophical Magazine*, Vol. 1, No. 5, 1956, pp. 450–458.

[7] Nishinaga, T., "Progress in Art and Science of Crystal Growth and Its Impacts on Modern Society," *Japanese Journal of Applied Physics*, Vol. 54, 2015, pp. 050101-1–050101-12.

[8] Burton, W. K., N. Cabrera, and F. C. Frank, "The Growth of Crystals and the Equilibrium Structure of Their Surfaces," *Proceedings of the Royal Society of London A*, Vol. 243, No. 866, 1951, pp. 299–358.

[9] Hirth, J. P., and G. M. Pound, *Condensation and Evaporation, Nucleation and Growth Kinetics*, Oxford: Pergamon, 1963, Chapter D.

[10] Abbink, H. C., R. M. Broudy, and G. P. McCathy, "Surface Processes in the Growth of Silicon on (111) Silicon in Ultrahigh Vacuum," *Journal of Applied Physics*, Vol. 39, No. 10, 1968, pp. 4673–4681.

[11] Kasper, E., "Growth Kinetics of Si-Molecular Beam Epitaxy," *Applied Physics A*, Vol. 28, No. 2, 1982, pp. 129–135.

[12] Nishinaga, T., and K.-I. Cho, "Theoretical Study of Mode Transition Between 2d-Nucleation and Step Flow in MBE of GaAs," *Japanese Journal of Applied Physics*, Vol. 27, No. 1, 1988, pp. L12–L14.

[13] Nishinaga, T., and T. Suzuki, "The Role of Step Kinetics in MBE of Compound Semiconductors," *Journal of Crystal Growth*, Vol. 115, No. 1–4, 1991, pp. 395–405.

[14] Mochizuki, K., and T. Nishinaga, "MBE Growth of $GaAs_{1-x}Sb_x$ and $In_yGa_{1-y}As$ and Application of BCF Theory to Study the Alloy Composition," *Japanese Journal of Applied Physics*, Vol. 27, No. 9, 1988, pp. 1585–1592.

[15] Kimoto, T., and H. Matsunami, "Surface Kinetics of Adatoms in Vapor Phase Epitaxial Growth of SiC on 6H-SiC{0001} Vicinal Surfaces," *Journal of Applied Physics*, Vol. 75, No. 2, 1994, pp. 850–859.

[16] Mochizuki, K., "Theoretical Consideration of Step-Flow and Two-Dimensional Nucleation Modes in Homoepitaxial Growth of 4H-SiC on (0001) Vicinal Surfaces Under Silicon-Rich Condition," *Applied Physics Letters*, Vol. 93, 2008, pp. 222108-1–222108-3.

[17] Burk, A., and L. B. Rowland, "Homoeptaxial VPE Growth of SiC Active Layers," *Physica Status Solidi B*, Vol. 202, No. 1, 1997, pp. 263–279.

[18] Rupp, R., et al., "Silicon Carbide Epitaxy in a Vertical CVD Reactor: Experimental Results and Numerical Process Simulation," *Physica Status Solidi B*, Vol. 202, No. 1, 1997, pp. 281–304.

[19] Kordina, O., et al., "Growth of SiC by "Hot-Wall" CVD and HTCVD," *Physica Status Solidi B*, Vol. 202, No. 1, 1997, pp. 321–334.

[20] Nishizawa, S., and M. Pons, "Growth and Doping Modeling of SiC-CVD in a Horizontal Hot-Wall Reactor," *Chemical Vapor Deposition*, Vol. 12, No. 8–9, 2006, pp. 516–522.

[21] Kuroda, N., et al., "Step-Controlled VPE Growth of SiC Single Crystals at Low Temperatures," *Solid State Devices and Materials*, Tokyo, Aug. 25–27, 1987, pp. 227–230.

[22] Kong, H. S., et al., "Growth, Doping, Device Development and Characterization of CVD Beta-SiC Epilayers on Si(100) and alpha-SiC(0001)," *MRS Proceedings*, Vol. 97, 1987, pp. 233–246.

[23] Kong, H. S., J. T. Glass, and R. F. Davis, "Chemical Vapor Deposition and Characterization of 6H-SiC Thin Films on Off-Axis 6H-SiC Substrates," *Journal of Applied Physics*, Vol. 64, No. 5, 1988, pp. 2672–2679.

[24] Itoh, A., et al., "High-Quality 4H-SiC Homoepitaxial Layers Grown By Step-Controlled Epitaxy," *Applied Physics Letters*, Vol. 65, No. 11, 1994, pp. 1400–1402.

[25] Kimoto, T., et al., "Step Bunching Mechanism in Chemical Vapor Deposition of 6H- and 4H-SiC{0001}," *Journal of Applied Physics*, Vol. 81, No. 8, 1997, pp. 3494–3500.

[26] Meziere, M., et al., "Modeling and Simulation of SiC CVD in the Horizontal Hot-Wall Reactor Concept," *Journal of Crystal Growth*, Vol. 267, No. 3–4, 2004, pp. 436–451.

[27] Saito, H., and T. Kimoto, "4H-SiC Epitaxial Growth on SiC Substrates with Various Off-Angles," *Materials Science Forum*, Vol. 483–485, 2005, pp. 89–92.

[28] Pearson, E., et al., "Computer Modeling of Si and SiC Surfaces and Surface Processes Relevant to Growth from the Vapor," *Journal of Crystal Growth*, Vol. 70, No. 1–2, 1984, pp. 33–40.

[29] Fujihira, T., "Theory of Semiconductor Superjunction Devices," *Japanese Journal of Applied Physics*, Vol. 36, No. 10, 1997, pp. 6254–6262.

[30] Kosugi, R., et al., "First Experimental Demonstration of SiC Superjunction (SJ) Structure by Multiepitaxial Method," *International Symposium on Power Semiconductor Devices and ICs*, Waikoloa, June 2014, pp. 346–349.

[31] Lorenz, L., et al., "COOLMOS—A New Milestone in High Voltage Power MOS," *International Symposium on Power Semiconductor Devices and ICs*, Toronto, May 1999, pp. 3–10.

[32] Kosugi, R., "Development of SiC Superjunction (SJ) Device by Deep Trench-Filling Epitaxial Growth," *Materials Science Forum*, Vol. 740–742, 2013, pp. 785–788.

[33] Kojima, K., et al., "Filling of Deep Trench by Epitaxial SiC Growth," *Materials Science Forum*, Vol. 740–742, 2013, pp. 793–796.

[34] Takeuchi, Y., et al., "SiC Migration Enhanced Embedded Epitaxial (ME^3) Growth Technology," *Materials Science Forum*, Vol. 527–529, 2006, pp. 251–254.

[35] Sugiyama, N., et al., "Growth Mechanism and 2d Aluminum Dopant Distribution of Embedded Trench 4H-SiC Region," *Materials Science Forum*, Vol. 600–603, 2009, pp. 171–174.

[36] Schöner, A., et al., "In Situ Nitrogen and Aluminum Doping in Migration Enhanced Embedded Epitaxial Growth of 4H-SiC," *Materials Science Forum*, Vol. 600–603, 2008, pp. 175–178.

[37] Negoro, Y., et al., "Embedded Epitaxial Growth of 4H-SiC on Trenched Substrates and pn Junction Characteristics," *Microelectronics Engineering*, Vol. 83, No. 1, 2006, pp. 27–29.

[38] Ji, S. Y., et al., "Influence of Growth Pressure on Filling 4H-SiC Trenches by CVD Method," *Japanese Journal of Applied Physics*, Vol.55, 2016, pp. 01AC04-1–01AC04-4.

[39] Mochizuki, K., et al., "Analysis of Trench-Filling Epitaxial Growth of 4H-SiC Based on Continuous Fluid Approximation Including Gibbs–Thomson Effect," *Materials Science Forum*, Vol. 897, 2017, pp. 47–50.

[40] Krishnamachari, B., "Gibbs–Thomson Formula for Small Island Sizes: corrections for High Vapor Densities," *Physical Review B*, Vol. 54, No. 12, 1996, pp. 8899–8907.

[41] Mochizuki, K., et al., "Numerical Analysis of the Gibbs–Thomson Effect on Trench-Filling Epitaxial Growth of 4H-SiC," *Applied Physics Express*, Vol. 9, 2016, pp. 03560-1–03560-4.

[42] McDonald, J. E., "Homogeneous Nucleation of Supercooled Water Drops," *Journal of Meteorology*, Vol. 10, 1953, pp. 416–433.

[43] J., S. Y., et al., "Filling 4H-SiC Trench Towards Selective Epitaxial Growth by Adding HCl to CVD Process," *Applied Physics Express*, Vol. 8, 2015, pp. 065502-1–065502-4.

[44] Yamauchi, S., et al., "200 V Super Junction MOSFET Fabricated by High Aspect Ratio Trench Filling," *International Symposium on Power Semiconductor Devices and ICs*, Naples, June 2006, pp. 1–4.

[45] Hara, K., et al., "150 mm Silicon Carbide Selective Embedded Epitaxial Growth Technology by CVD," *Materials Science Forum*, Vol. 897, 2017, pp. 43–46.

[46] Mochizuki, K., et al., "First topography simulation of SiC-chemical-vapor-deposition trench filling, demonstrating the essential impact of the Gibbs–

Thomson effect," *International Electron Devices Meeting*, San Francisco, December 4–6, 2017, pp. 788–791.

[47] Kosugi, R., et al., "Strong Impact of Slight Trench Direction Misalignment from [11$\bar{2}$0] on Deep Trench Filling Epitaxy for SiC Superjunction Devices," *Japanese Journal of Applied Physics*, Vol. 56, 2017, pp. 04CR05-1–04CR05-4.

第 7 章

制作工艺

7.1 引言

本章将介绍制作 GaN 和 4H-SiC 功率器件的基本工艺步骤,但不包括已在第 6 章中介绍过的外延生长工艺。7.3 节和 7.4 节将分别介绍 4H-SiC 功率器件的离子注入和扩散工艺。7.2 节和 7.5~7.7 节将分别介绍 GaN 和 4H-SiC 功率器件的刻蚀、氧化、金属化和钝化工艺,并对两种器件工艺进行对比。

7.2 刻蚀

对于传统湿法化学刻蚀而言,GaN 和 4H-SiC 属于强惰性材料,因此通常使用高密度等离子体干法刻蚀工艺进行制作,如磁控反应离子、电子回旋共振等离子体或电感耦合等离子体(ICP)工艺,其中 ICP 的刻蚀速率最高。例如,首次使用铝/氢/氩等离子体进行的 GaN ICP 工艺,其刻蚀速率便达到了 $0.7\mu m/min$[1,2]。在线圈上施加 RF(射频)功率可以形成 ICP 等离子体(见图 7.1)。通过对离子能和等离子体密度进行解耦,即使在离子能和电子能较低的情况下也可以实现均匀的密度和能量分布。因此 ICP 刻蚀可以在保持高刻蚀速率的前提下产生较低损伤。通过向刻蚀样品施加 RF 偏压可以实现各向异性刻蚀。依据已报道的研究结果,本节将论述 GaN 和 4H-SiC 的 ICP 刻蚀及湿法化学刻蚀工艺。

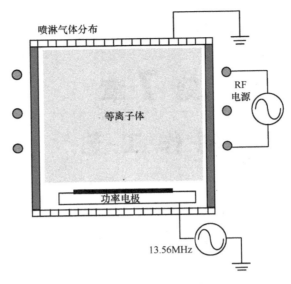

图 7.1 ICP 刻蚀机示意图

7.2.1 ICP 刻蚀

GaN 的 ICP 刻蚀通常采用卤素和甲烷基 ICP 等离子体工艺[3-8]。Ⅲ 族卤素刻蚀产品的挥发性限制了卤素基等离子体的刻蚀速率。在卤素基等离子体刻蚀中,氯基等离子体的刻蚀速率比较高,而甲烷基等离子体的刻蚀速率比较低。虽然上述 ICP 刻蚀技术可将等离子体损伤降至最低,但对器件特性的影响却不容忽视。根据 Ohta 等人的报道,该类损伤可以通过 850℃ 热处理得以消除[9]。

SiC 的 ICP 刻蚀通常采用氟基、氯基和溴基等离子体工艺[10-14]。第 6 章、第 8 章和第 9 章还介绍了用于制作超结结构的 25μm 深 4H-SiC 沟槽工艺[15]。

7.2.2 湿法化学刻蚀

半导体的湿法化学刻蚀通常包含半导体表面的氧化及其所产生氧化物后续的分解过程。氧化所需的空穴可以通过化学反应或电化学电路提供[16]。

7.2.2.1 化学刻蚀

在 GaN 化学刻蚀工艺中,镓极性表面和氮极性表面分别显示出不同的刻蚀特性;在 GaN 外延层[17]和 GaN 体材料[18]中,只有氮极性表面在氢氧化钾(KOH)水溶液中可以刻蚀,氮极性表面的刻蚀机理通过下式表达[19]:

$$2GaN+3H_2O \rightarrow Ga_2O_3+2NH_3 \tag{7.1}$$

KOH 是 Ga_2O_3 的催化剂和溶解剂。与氮极性表面不同，镓极性（0001）表面通常在热磷酸（470~490K）中进行选择性侵蚀，形成六边形腐蚀坑[20-24]。

而 SiC 在磷酸中的化学刻蚀速率过低，因此不经常使用，取而代之的是使用熔融的 KOH 来氧化 SiC，并去除形成的氧化物[25]。使用 KOH 必须注意避免钾污染，特别是 SiC 的氧化过程更应引起注意[26]。此外由于 KOH 使位错与表面交界处的刻蚀速率增快，因此需注意可能出现的位错坑。

在沟槽结构制作完成后所进行的湿法刻蚀工艺中，据报道在 358K 温度下使用 25% 的四甲基氢氧化铵（TMAH）可以有效地获得光滑的 GaN（1100）平面[27]。TMAH 是一种用于酸性光刻胶显影的碱性溶剂，广泛应用于光刻工艺，且不受碱金属的污染。

7.2.2.2 光电化学刻蚀

光电化学刻蚀（PEC）与 GaN 晶体的极性无关[28]。首次 GaN PEC 是在 KOH 或 HCl 水溶液中使用波长为 325nm 的氦-镉激光器实现的，刻蚀速率分别为 400nm/min 和 40nm/min[29]。PEC 还可用于制作 AlGaN/GaN 异质结构，比如 $Al_{0.2}Ga_{0.8}N$/GaN 异质结场效应晶体管（HFET）的湿法栅极凹槽就是使用 35mW/cm^2 汞弧光灯和 0.5mol% KOH 制作的，这一过程需要在 2in 晶圆边缘区域的台面刻蚀表面与铂栅极之间施加 2V 的 DC 偏压（见图 7.2）[30]。

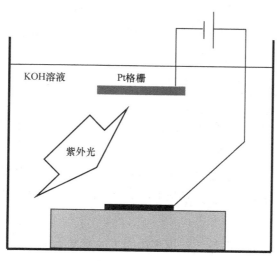

图 7.2 用于刻蚀 AlGaN/GaN 异质结构的 PEC 刻蚀工具示意图[30]

在 SiC p-n 结的制作过程中，由于 n 型 SiC 中缺少空穴，p 型 SiC 的选择性刻蚀可以不使用紫外光照射[31,32]，而 n 型 SiC 的刻蚀可以在光能大于其带隙能量的紫外光照射下制作[33-37]。光生空穴在电解质（如 KOH 溶液）/n 型 SiC 界面处积累，而空穴在电解质/p 型 SiC 界面处耗尽。

7.3 离子注入

如 3.1 节和 3.7 节所述，GaN 和 4H-SiC 功率器件通常采用 n 型原始材料制作，p^+n 二极管的有源区应避免使用 p 型掺杂剂注入。在单极功率器件中，通常使用铝离子注入 4H-SiC 的方法；但是极少使用镁离子注入 GaN，因为镁离子的激活率很低（见 3.7 节）。也不常用硼离子注入 4H-SiC 的方式，因为高能离子穿过 4H-SiC（或 GaN）晶体发生连续碰撞，导致形成硅空位 V_{Si} 和碳空位 V_C（或镓空位 V_{Ga} 和氮空位 V_N），以及硅间隙原子 I_{Si} 和碳间隙原子 I_C（或镓间隙原子 I_{Ga} 和氮间隙原子 I_N）硼离子注入后需要进行退火处理，在退火过程中，SiC 中的硼会重新分布并变得很大（见 6.6 节和 7.4 节）。

离子注入后再退火过程中会出现表面退化，所以在 1500K 高温条件下对离子注入后的 GaN 进行退火时通常会使用 SiN 帽层进行保护[38]，在 1950K 高温下对离子注入后的 SiC 进行退火时通常会使用碳帽进行保护[39]。

n 型掺杂剂注入通常使用硅离子注入 GaN，以及氮离子和磷离子注入 4H-SiC 的方法。

7.3.1 离子注入 GaN

在孤立 GaN 衬底上生长的高质量 n 型 GaN 外延层中尝试进行镁离子注入，以下三项实验结果证实 GaN 转换成了 p 型材料[40]：

1）GaN 外延层的低温光致发光谱显示了与镁受主相关的发光性能。

2）垂直型二极管显示了清晰的 *I-V* 整流特性，并在正向偏置条件下发出紫外光和蓝绿光。

3）观测到正的霍尔系数，但提高镁受主的激活率仍是一项挑战。

使用硅离子注入法制作 AlGaN/GaN 高电子迁移率晶体管（HEMT）[41]可以降低金属/GaN 接触电阻，实现轻掺杂沟道区。硅离子注入 p 型 GaN 可用于制作阈值电压为+3.4V 的常关型 GaN 金属-绝缘体-半导体场效应晶体管（MIS-FET）[42]。

7.3.2 铝离子注入 4H-SiC

对于硼离子注入硅而言，采用大倾角（如与硅[100]晶面形成 7°夹角）注入随机表层的过程中，可产生沟道散射效应[43]，或称为反常剖面展宽[44]。沟道散射效应是指离子穿过表面层后，一些高能离子由随机移动散射成沿沟道方向移动，并深深穿透到无损的背面晶体中。当注入倾角为 0°时，准直离子束受非晶氧化层散射出现非晶化抑制沟道效应，注入的深度随表面氧化层厚度的增加而减少，当采用特定注入能量并以更大倾角注入时，会出现沟道散射效应，注入的深度随氧化层厚度的增加而增加[45]。

对于铝离子注入 4H-SiC 而言，衬底通常与（0001）晶面存在 4°~8°的偏离，以便实现阶梯流动式外延生长[46]。由于存在铝沟道散射效应，表面氧化物对沟道效应的影响可能变得更大。此外，由于 4H-SiC 中铝的扩散率可忽略不计，使用多能量铝注入一定会形成盒状分布。注入能变化的顺序（升序或降序）可以确定究竟是沟道散射效应还是非晶化抑制沟道效应对盒状分布起着主导作用。Mochizuki 等人在室温下对带有和不带有 35nm 厚 SiO_2 层的 4H-SiC 晶圆（从（0001）向[1120]偏离 8°）进行五层铝注入，获得了 0.3μm 深的盒状分布，平均的平台浓度为 $1×10^{19} cm^{-3}$ [47]。注入能（keV）及其相应剂量（$×10^{13} cm^{-2}$）分别为 220/10、160/5、110/7、70/6 和 35/3。当注入浓度能不断下降时，不带 SiO_2 层的铝离子注入浓度拖尾延伸较短，带有 SiO_2 层的注入浓度拖尾比不带 SiO_2 层的延伸要长，这种差异是由沟道散射效应造成的。相反，当注入能不断增大时，带有和不带有 SiO_2 层的注入浓度拖尾只存在微小的差别。由此可见，非晶化抑制沟道效应比沟道散射效应对盒状分布的影响要弱。

使用皮尔逊频率分布函数[48]成功地对 4H-SiC 中的注入离子分布的广泛选择性进行了分析[49]。但是对于铝之类的重离子来说，注入分布的大沟道拖尾却与单皮尔逊分布函数的计算结果存在一定的偏差[49-51]。双皮尔逊分布函数是两个皮尔逊分布函数的加权之和[48]，用于对注入分布的随机散射部分和沟道部分建模[52]。铝离子的深度分布 $N(x)$ 通过下式表示

$$N(x)=D_1 f_1(x)+D_2 f_2(x) \tag{7.2}$$

$$f_i(x) = K_i \left[1 + \{(x-R_{pi})/A_i - n_i/r_i\}^2\right]^{-m_i} \exp\left[-n_i \arctan\{(x-R_{pi})/A_i - n_i/r_i\}\right]$$
$$(i=1, 2) \tag{7.3}$$

式中，f_1 和 f_2 分别为注入分布的随机散射部分和沟道部分的皮尔逊Ⅳ型分布函数；D_1 和 D_2 分别为各自皮尔逊分布函数对应的剂量值。在皮尔逊Ⅳ型分布函数中，K_1 和 K_2 是常量，R_{p1} 和 R_{p2} 为投影射程，n_1、n_2、r_1、r_2、A_1、A_2、m_1 和 m_2 是与射程离散 ΔR_{p1} 和 ΔR_{p2}、偏度 γ_1 和 γ_2，以及峰度 β_1 和 β_2 相关的参数，具体关系如下

$$r_i = -(2 + 1/b_{2i}) \tag{7.4a}$$

$$n_i = -r_i b_{1i}/(4b_{0i}b_{2i} - b_{1i}^2)^{0.5} \tag{7.4b}$$

$$m_i = -1/(2b_{2i}) \tag{7.4c}$$

$$A_i = m_i r_i b_{1i}/n_i \tag{7.4d}$$

$$b_{0i} = \Delta R_{pi}^2 (4\beta_i - 3\gamma_i^2) C \tag{7.4e}$$

$$b_{1i} = -\gamma_i \Delta R_{pi}(\beta_i + 3) C \tag{7.4f}$$

$$b_{2i} = -(2\beta_i - 3\gamma_i^2 - 6) C \tag{7.4g}$$

$$C = 1/[2(5\beta_i - 6\gamma_i^2 - 9)] \quad (i=1,2) \tag{7.4h}$$

剂量比 R 定义为

$$R = D_1/(D_1 + D_2) \tag{7.5}$$

为了避免 R_{p2} 的任意性[53]，将其设置为与 R_{p1} 相等（见图 7.3）。

通过对双皮尔逊参数与单皮尔逊参数的对比显示，不带 SiO_2 层离子注入的 R_p、ΔR_{p1} 和 γ_1 的关系几乎与文献 [49, 50] 中所述相同，但与文献 [51] 中所述略有不同。尽管单皮尔逊模型中的 β 值仍无法避免其复杂性，但图 7.3e 中获得的 β_1 和 γ_1 之间的关系与式（7.6c）中的关系极为相近[51]。

$$\beta_1 = 1.19\beta_{1o} \tag{7.6a}$$

$$\beta_{1o} = [39\gamma_1^2 + 48 + 6(\gamma_1^2 + 4)^{1.5}]/(32 - \gamma_1^2) \tag{7.6b}$$

$$\beta = 1.30\beta_o \tag{7.6c}$$

图 7.3a 显示，带有 SiO_2 层比不带 SiO_2 层离子注入的 R_p 值要小，这与注入过程中存在 SiO_2 层有关。图 7.3b 显示，带有 SiO_2 层比不带 SiO_2 层离子注入的 R 值要小，这是由于沟道散射效应产生了更多的铝离子沟道造成的。对于带有 35nm 厚 SiO_2 层的铝离子注入来说，其 R 值单调递增的条件为[54]：

- 剂量为 $1\times10^{14} cm^{-2}$，注入能量大于 300keV。
- 剂量为 $1\times10^{15} cm^{-2}$ 或以上，注入能量为 35keV 或以上。

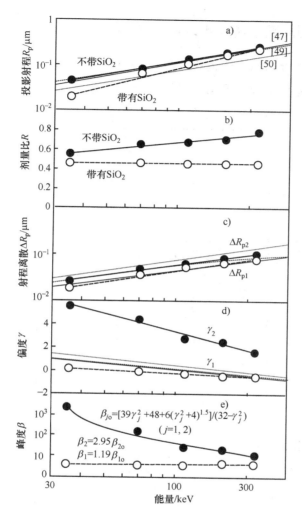

图 7.3 双皮尔逊参数与注入能量之间的函数关系
（与之前报道的投影射程、射程离散和偏度进行了对比）

如上所述，以上任一条件下非晶化抑制沟道效应的作用[43]都可能增大。

横向射程离散可以通过横向浓度分布得出，横向浓度分布是一维双皮尔逊分布函数乘以另一个分布函数（例如高斯函数）得出的。该二维模型有助于有效模拟 4H-SiC 功率器件的电流-电压特性[55,56]。除双皮尔逊方法外，还采用了二元碰撞近似方法（BCA）[57,58]的蒙特卡洛模拟。BCA 模拟可以得出由 SiC 衬底的表面晶向偏离导致的不对称横向离散[59]。

7.3.3 氮离子和磷离子注入 4H-SiC

采用单皮尔逊方法分析了注入的氮离子和磷离子的浓度分布[49,60-62]。图7.4中的 R_p 和 ΔR_p 可表示为氮离子和磷离子注入 SiC 时注入能量的函数。氮原子和磷原子在注入后退火过程中都产生了极微小的扩散[63]。

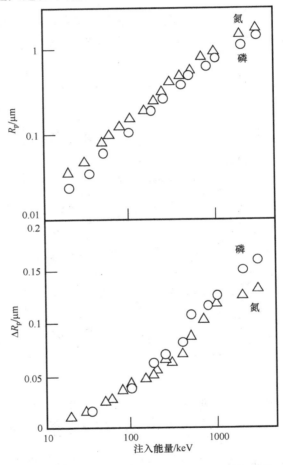

图 7.4 氮离子和磷离子注入 SiC 中，注入能量与射程 R_p 和射程离散 ΔR_p 的关系[49,60-62]

7.4 扩散

将硅原子和镁原子注入 GaN 中，然后在 1450℃ 下退火[64]，将铝原子注入

4H-SiC 中,然后在 1700℃下退火[65],通过二次离子质谱法(SIMS)可观察到这些原子产生了极微小的扩散。与上述离子注入相比,硼在 SiC 中扩散的机理非常复杂。然而,由于最近有报道显示利用硼扩散来制作 4H-SiC 金属-绝缘体-半导体场效应晶体管可降低 4H-SiC 导带边缘附近的界面态密度[66],因此理解这一机理变得更为重要。

7.4.1 SiC 中硼扩散的历史背景

对 SiC 中硼扩散的最初分析是基于 6H-SiC 的硼空位模型[67]。然而,对氮掺杂的 4H-SiC 和铝掺杂的 6H-SiC 中硼浓度分布的详细分析表明,I_{Si}(而非 V_{Si})控制着硼的扩散[68]。鉴于 I_C 的参与,需要重新分析 I_{Si} 介导的硼扩散[69]。依据对富同位素 4H-SiC 进行的自扩散实验,I_{Si} 和 I_C 的扩散率处于相同的数量级,并且在特定的实验条件下,任一缺陷都与硼的首选晶格位点密切相关。对 3C-SiC 晶体结构的理论计算[70-72]也表明 I_{Si} 和 I_C 比 V_{Si} 和 V_C 的移动性强得多。假设 I_{Si} 和 I_C 在 4H-SiC 中具有相同的迁移率,可以通过 I_{Si} 和 I_C 根据点缺陷的某一初始分布对硼扩散进行模拟。

SiC 中与硼相关的杂质中心具有两个关键特征:电离能约为 0.30eV 的浅受主和电离能约为 0.65eV 的深能级[73]。虽然浅受主缺陷的形成是在硅位点处(B_{Si})出现了偏离中心位置的替位硼原子[74]导致的,但与硼相关深能级的产生原因仍不明确。从头计算(ab initio,全电子的非经验计算)方法推翻了 B_{Si}-V_C 缺陷[73],所以 B_{Si}-Si_C 复合缺陷是可能的原因[75]。此外,碳原子位点上的替位硼原子(B_C)和 B_C-C_{Si} 复合缺陷等都是可能的原因[76]。富硅条件下硼掺杂 4H-SiC 的同质外延生长更有利于硼相关深能级中心的形成[77],然而相似实验表明富碳条件有利于硼浅受主能级的产生[69]。这些不同的位点竞争现象表明,硼原子可以同时占据与硅和碳相关的位点,因此可以推测硼相关深能级原因在于 B_C 的观点[69]。根据 3C-SiC 的理论[70,71],硼缺陷的迁移会形成硼间隙(I_B)原子而不是硼间隙原子对,而硼缺陷的存在影响了硼在硅中的扩散[78,79]。虽然通过 I_B 扩散模拟来计算硼浓度分布是理想的方法,但仍不清楚与 4H-SiC 中 I_B 最相关的因素。为了通过实验重现获得硼扩散分布来设计 4H-SiC 功率器件,应用了商业制程模拟器中的硼-间隙原子对扩散模型[57],该模型主要是为硅应用而优化设计的。在 7.4.2 节和 7.4.3 节中引用了文献 [80,81] 中报道的 4H-SiC 中的硼浓度分布,因为 500℃高温注入(200keV/$4×10^{14}cm^{-2}$)的硼离子的退火条件发生了系统性的变化。

7.4.2 双子晶格扩散建模

假设 I_{Si} 和 I_C 在本身子晶格上的扩散[82]与3C-SiC中硼扩散[83]的理论计算结果一致,则通过硅位点和碳位点上的硼原子(B_{Si} 和 B_C)的替代掺杂形成 I_B,I_B 表示为

$$B_{Si}+I_{Si} \leftrightarrow I_B (\text{Ⅰ型}) \quad (7.7)$$

$$B_C+I_C \leftrightarrow I_B (\text{Ⅱ型}) \quad (7.8)$$

式中省略了电荷状态的表达式。

在3C-SiC情况下,I_B(Ⅰ型)和 I_B(Ⅱ型)是碳配位的四面体位点、六方体位点或者是硅位点或碳位点的分裂间隙[71]。式(7.7)和式(7.8)给出的反应与硼间隙原子对扩散模型[84]的如下反应相对应:

$$B_{Si}^{j}+I_{Si}^{m} \leftrightarrow (B_{Si}I_{Si})^{u}+(j+m-u)h^{+} \quad (7.7a)$$

$$B_C^{k}+I_C^{n} \leftrightarrow (B_C I_C)^{v}+(k+n-v)h^{+} \quad (7.8a)$$

式中,电荷状态 j,k,m,n,u,$v \in \{0, \pm 1, \pm 2, \cdots\}$ 和空穴 h^+。根据先前文献[71]中的计算,可以将3C-SiC中的 I_{Si} 从不带电状态充电到+4,将 I_C 从-2充电到+2。

假设在4H-SiC中 I_{Si} 和 I_C 的电荷状态变化与在3C-SiC中的相同,则式(7.7a)和式(7.8a)中 m 和 n 的范围限制为 $m \in \{0, 1, 2, 3, 4\}$ 和 $n \in \{0, \pm 1, \pm 2\}$。

图7.5显示了掩埋硼掺杂层外延生长的4H-SiC结构中硼的扩散[85]建模。在这种条件下,点缺陷的浓度是热力学平衡状态下的浓度。因此可以使用费米模型分析,在该模型中,点缺陷对掺杂物扩散的所有影响都可以用原子对扩散系数来表示[86]。在本例中,式(7.7a)和式(7.8a)中原子对的扩散系数是 $(B_{Si}I_{Si})^u$ 和 $(B_C I_C)^v$。通常,当掺杂浓度超过本征载流子浓度 n_i 时[87],如下式所示:

$$n_i(T) = 1.70 \times 10^{16} T^{1.5} \exp(-2.08 \times 10^4/T)(\text{cm}^{-3}) \quad (7.9)$$

扩散与浓度相关[86]。因此,原子对扩散率 D(杂质A和填隙I)可表示为

$$D_{AI}=D_{AI}^{0}+D_{AI}^{+}(p/n_i)^{+1}+D_{AI}^{++}(p/n_i)^{+2}+D_{AI}^{-}(p/n_i)^{-1}+D_{AI}^{=}(p/n_i)^{-2} \quad (7.10)$$

式中,p 为空穴浓度;上标"++"和"="代表+2和-2。

如7.4.1节所述,当SiC结构在富碳条件下生长时,把硼原子作为浅受主(B_{Si}^-)引入硅位点。因此式(7.7a)和式(7.10)变为

$$B_{Si}^- + I_{Si}^m \leftrightarrow (B_{Si}I_{Si})^u + (-1+m-u)h^+ \quad (7.11)$$

$$D(B_{Si}I_{Si}) = D(B_{Si}I_{Si})^0 + D(B_{Si}I_{Si})^+(p/n_i)^{+1} + D(B_{Si}I_{Si})^{++}(p/n_i)^{+2} + D(B_{Si}I_{Si})^-(p/n_i)^- +$$
$$D(B_{Si}I_{Si})^=(p/n_i)^2 \quad (7.12)$$

式中，$m \in \{0, 1, 2, 3, 4\}$；$u \in \{0, \pm1, \pm2\}$。

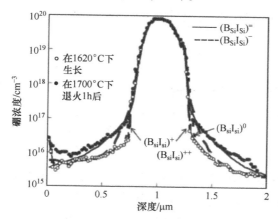

图 7.5 在退火前（空心圆）和 1700°C 退火 1h 之后，带有掩埋硼掺杂层的外延生长 4H-SiC 结构中的硼浓度分布[85]。以双负电荷 $B_{Si}I_{Si}$ 对的扩散系数 $1\times10^{-15} cm^2/s$ 为参数而模拟得到的分布（实曲线）可以精确地再现实验结果（实心圆）

假设式（7.12）右边的单项可以主导（$B_{Si}I_{Si}$）对的扩散。在图 7.5 中，使用下列 5 个（$B_{Si}I_{Si}$）对的扩散系数的其中一个来拟合 1700°C 下退火 1h 后的分布：中性无电荷对、单正电荷对、双正电荷对、单负电荷对和双负电荷对。如图 7.5 所示，以双负电荷 $B_{Si}I_{Si}$ 对的扩散系数 $1\times10^{-15} cm^2/s$ 为参数模拟得到的分布可以精确地再现实验得出的浓度分布，而使用其他四个扩散系数模拟出的分布则与实验结果不符。因此选取（$B_{Si}I_{Si}$）$^=$ 来模拟 B_{Si}^- 扩散。

下一步对 B_C 的扩散[71]进行建模。将 B_C 看作受主，式（7.8a）变为

$$B_C^- + I_C^n \leftrightarrow (B_C I_C)^v + (-1+n-v)h^+ \quad (7.13)$$

式中，n 和 $v \in \{0, \pm1, \pm2\}$，而式（7.10）变为

$$D(B_C I_C) = D(B_C I_C)^0 + D(B_C I_C)^+(p/n_i)^{+1} + D(B_C I_C)^{++}(p/n_i)^{+2} + D(B_C I_C)^-(p/n_i)^{-1} +$$
$$D(B_C I_C)^=(p/n_i)^{-2} \quad (7.14)$$

在 p 型 6H-SiC 中，当硼的蒸气压较低时，硼内部扩散的扩散系数与 p 成正比[68]。假设在 4H-SiC 中存在相似的比例关系，7.4.3 节使用单正电荷对 $(B_C I_C)^+$ 的扩散系数可以来模拟 B_C^- 的扩散。

7.4.3 半原子模拟

根据蒙特卡洛模拟得到的初始点缺陷分布计算了硼离子注入的扩散[82]。在计算中，连续性方程为

$$\partial/\partial t(C_I + C_{(B_{Si}I)^=} + C_{(B_CI)^+}) = -\text{div}(J_I + J_{(B_{Si}I)^=} + J_{(B_CI)^+}) - K_r(C_I C_V - C_I^* C_V^*)$$
(7.15)

上式可以由

$$J_{(B_{Si}I)^=} = -D_{(B_{Si}I)^=}\{-\text{grad}[C_{B_{Si}}^-(C_I/C_I^*) + C_{B_{Si}}^-(C_I/C_I^*)(qE/kT)]\} \quad (7.16)$$

和

$$J_{(B_CI)^+} = -D_{(B_CI)^+}\{-\text{grad}[C_{B_C}^-(C_I/C_I^*) + C_{B_C}^-(C_I/C_I^*)(qE/kT)]\} \quad (7.17)$$

求解得到。

式中，C_I 和 C_V 分别为间隙原子和空位浓度；C_I^* 和 C_V^* 分别为平衡状态下的间隙原子和空位浓度；J_I 为间隙通量；K_r 为间隙-空位体复合系数；q 为电荷；E 为电场；k 为玻尔兹曼常数；T 为绝对温度。

如式 (7.16) 和式 (7.17) 所示，$(B_{Si}I)^=$ 和 $(B_CI)^+$ 的通量都考虑了电场的影响。模拟的第一步是获得初始注入的硼分布以及点缺陷的初始分布。一旦创建了 I_{Si} 和 I_C，就将它们视为相同的 I（来源未知）。同样地，将创建的 V_{Si} 和 V_C 视为相同的 V。因此，式 (7.12) 和式 (7.14) 可简化为

$$D_{(B_{Si}I)} = D_{(B_{Si}I)^=}(p/n_i)^{-2} \quad (7.18)$$

$$D_{(B_CI)} = D_{(B_CI)^+}(p/n_i)^{+1} \quad (7.19)$$

在蒙特卡洛模拟中，假设离子注入方向由 4H-SiC 的表面从 (0001) 向 [1120] 偏离了 8°，且设定硼离子束发散度为 0.1°。注入的硼离子占据硅位点 (r_{Si}) 或碳位点 (r_C) 的概率按如下规定：在 200keV、500℃、硼离子注入为 $4\times10^{14}\text{cm}^{-2}$，且假定 $r_{Si} = r_C = 0.5$ 的试验条件下[81,82]，计算得出 B_{Si}^-、B_C^-、I 和 V 的注入浓度分布。其他温度相关参数的详细确定过程见参考文献 [82]。

如图 7.6 所示，根据双子晶格建模（见 7.4.2 节）模拟的硼浓度图，准确地展示了 $T=1900℃$、$t=15\text{min}$ 条件下本底掺杂能级 (N_b) 从 n 型到 p 型的测量分布拖尾区域（符号表示）[81]。在 B_{Si}^- [$N_b = 1\times10^{19}\text{cm}^{-3}$（n 型）]、$B_C^-$ [$N_b = 2\times10^{15}\text{cm}^{-3}$（n 型）和 $4\times10^{19}\text{cm}^{-3}$（p 型）] 的条件下，存在拖尾区。对扩散随时间而言，在 $N_b = 4\times10^{19}\text{cm}^{-3}$（p 型）、$T=1400℃$ 的条件下，硼浓度随退

火时间（5~90min）的变化[81]如图 7.7 所示。随时间变化的硼扩散分布（符号表示）可以根据以下参数得到精确再现，$D_{(B_{Si}I)}^{=} = 3 \times 10^{-18} \text{cm}^2/\text{s}$，$D_{(B_C I)}^{+} = 6 \times 10^{-12} \text{cm}^2/\text{s}$ 和 $K_r = 3 \times 10^{-16} \text{cm}^3/\text{s}$。

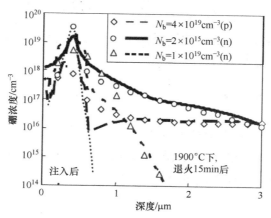

图 7.6 1900°C 退火 15min 后，在 4H-SiC 中测量
（符号表示[81]）和模拟得到的硼浓度分布

还应注意以下尚待解决的问题：为了应用开发的半原子模拟来拟合其他通过实验得到的硼浓度分布，必须优化 r_{Si}/r_C。由于这种优化与实验条件密切相关[69]，因此需要针对每种实验条件分别优化 r_{Si}/r_C。

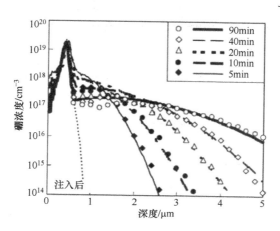

图 7.7 1400°C 退火后，在掺杂浓度为 $4 \times 10^{19} \text{cm}^{-3}$ 的 p 型 4H-SiC
掺杂中测得的（符号表示[81]）和模拟得到的硼浓度分布

7.5 氧化

对于硅来说,由于它具有优良的氧化物/半导体界面特性并且方便在制造器件中使用,因此热氧化是最有效的氧化物生长方法。但是,化合物半导体的热氧化比硅的热氧化复杂得多。

7.5.1 GaN 的热氧化

通过在 H_3PO_4 溶液中(pH 值为 3.5)氧化 GaN,利用氦-镉激光照明来生长厚度为 80nm 的 α-Ga_2O_3[88]。这种技术也用于制造 AlGaN/GaN 金属-绝缘体-半导体(MIS)高电子迁移率晶体管[89]。但是,在 Ga_2O_3 的五个相中,α 为亚稳相,β 为最稳相。

虽然将 GaN 在氧气环境、850°C 下氧化 12h 后得到了 β-Ga_2O_3[90],但其导带不连续能量 ΔE_C 非常小,仅为 0.1eV(见图 7.8[91]),限制了它用作 GaN MIS 场效应晶体管(MISFET)的栅极介电层。相较而言,GaN 上沉积 SiN_x 层的 ΔE_C 为 2.3eV,GaN 上沉积 Al_2O_3 层的 ΔE_C 为 2.1eV[92]。

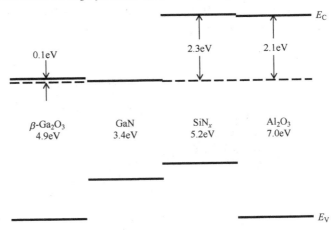

图 7.8 已报道的室温下的能带排列[91,92]

7.5.2 4H-SiC 的热氧化

SiC 的热氧化可表示为

$$2SiC+3O_2 \rightarrow 2SiO_2+2CO \tag{7.20}$$

理想情况下，CO 分子可从正在生长的氧化物中扩散出来。但是，一些碳原子可能会留在氧化物/SiC 界面的附近。尽管所谓的 Deal-Grove 模型[93,94]能够解释测得的氧化物厚度随氧化时间的变化关系，但测量结果与模型存在着偏差[95]。因此，尚未建立对热氧化的微观理解。据报道，热氧化物/4H-SiC 在 (0001) 晶面上的 ΔE_C（即 2.7~2.8eV）大于 (000$\overline{1}$) 晶面上的 ΔE_C (2.3~2.5eV)[96]。ΔE_C 的这种差异可能与氧化物和 4H-SiC 极性 (0001) 面的界面偶极子有关。

研究发现，与通过 N_2O 氧化和 O_2 氧化形成的氧化物相比，使用沉积在 4H-SiC 上并在 NO 或 N_2O 中氮化的氧化物更可靠[97]。据报道，氮在氧化物/SiC 界面处的堆积有助于改善界面质量[98]。

7.6 金属化

由于肖特基接触将在第 8 章中进行介绍，因此下面介绍 n 型和 p 型欧姆接触。

7.6.1 与 GaN 的欧姆接触

与 n-GaN 的欧姆接触通常采用钛/铝和钛/铝基多层结构（例如金/镍/铝/钛）[99]。用钛/铝在 900℃ 条件下退火 20s 可得到低于 $10^{-5}\Omega \cdot cm^2$ 的特征接触电阻 ρ_c[100]。在与 n^+GaN 的接触中，施主密度约为 $10^{19}cm^{-3}$，钨和硅化钨 (WSi_x) 导致了 ρ_c 较低 ($8 \times 10^{-5}\Omega \cdot cm^2$)[101]。

与 p-GaN 的欧姆接触中，即使在退火之前，钯和钯基接触（包括金/钯、金/铂/钯和金/钯/镁/钯）也表现出了欧姆特性[102]。这一发现可用"无序诱导间隙态（DIGS）模型"[103]进行解释，其中由于外延 Pd (111)/p-GaN (0001) 界面[104]中的应力而出现了费米能级钉扎效应。据报道，在流动的氮气环境中，金/钛/金/银/钯合金在 800℃ 条件下持续退火 1min 时，ρ_c 极低，仅为 $1 \times 10^{-6}\Omega \cdot cm^2$[105]。

7.6.2 与 4H-SiC 的欧姆接触

厚度为 50~100nm 的镍层与 4H-SiC ($\rho_c = 1 \times 10^{-6}\Omega \cdot cm^2$，施主密度为 $2 \times$

$10^{19} cm^{-3}$ 时）在约 1000°C 的温度下烧结可形成良好的 n 型欧姆接触[106,107]。据称，在金属/半导体界面附近积累的碳原子可降低 ρ_c[108]。然而，还未完全弄清其详细机理[107]。

对于与 4H-SiC 的 p 型欧姆接触，已了解铝基金属（例如，钛/铝和钛/镍/铝）的情况[109,110]。为了使 $\rho_c < 1 \times 10^{-5} \Omega \cdot cm^2$，当在 1000°C 条件下烧结持续 2min 时，受主浓度需超过 $3 \times 10^{19} cm^{-3}$[111]。为了避免使用低熔点（约 630°C）的铝，还对钯基[112]、镍基[113]和钛基的 p 型欧姆接触[114]进行了研究。

7.7 钝化

由于半导体的表面对高电场比较敏感，因此必须阻止表面原子自由结合。通常选用 SiO_2 作为 GaN 和 4H-SiC 功率器件的钝化层材料。

除了 SiO_2，高介电常数（高 k）介质膜已被用于 GaN 二极管。与使用相同厚度的 SiO_2 膜相比，高 k 介质膜产生的膨胀耗尽区更大，有望制作出具有更高击穿电压的器件。已使用 MOCVD[115-117]方法在台面型 GaN p^+n 二极管上形成了 SiO_2 和 CeO_2 的混合氧化物（k = 12.3），最终的击穿电压超过 2kV；而被 SiO_2（k = 3.9）钝化的二极管的击穿电压为 1kV[118]。

将高 k 介质膜应用于 4H-SiC 功率器件，对使用 HfO_2 的场板进行了仿真（参见第 11 章）[119]。仿真结果表明，高 k 场板可显著减轻电场的增强程度并增加击穿电压。

7.8 总结

在论述外延生长的第 6 章之后，本章介绍了 GaN 和 4H-SiC 功率器件的基本制造工艺。由于相关研究仍在积极进行中，因此读者还应参考有关 4H-SiC 氧化的最新论文。

参 考 文 献

[1] Shul, R. J., et al., "Comparison of Dry Etch Techniques for GaN," *Electronics Letters*, Vol. 32, No. 15, 1996, pp. 1408–1409.

[2] Shul, R. J., et al., "Inductively Coupled Plasma Etching of GaN," *Applied Physics Letters*, Vol. 69, No. 8, 1996, pp. 1119–1121.

[3] Lee, Y. H., et al., "Etch Characteristics of GaN Using Inductively Coupled Cl_2/Ar and Cl_2/BCl_3 Plasmas," *Journal of Vaccum Science and Technology A*, Vol. 16, No. 3, 1998, pp. 1478–1482.

[4] Kim, H. S., et al., "Effects of Inductively Coupled Plasma Conditions on the Etch Properties of GaN and Ohmic Contact Formations," *Materials Science and Engineering B*, Vol. 50, No. 1–3, 1997, pp. 82–87.

[5] Cho, H., et al., "Low Bias Dry Etching of III-Nitrides in Cl_2-Based Inductively Coupled Plasmas," *Journal of Electronics Materials*, Vol. 27, No. 4, 1998, pp. 166–170.

[6] Cho, H., et al., "Comparison of Inductively Coupled Plasma Cl_2 and Cl_4/H_2 Etching of III-Nitrides," *Journal of Vacuum Science and Technology A*, Vol. 16, No. 3, 1998, pp. 1631–1635.

[7] Lee, J. W., et al., "Dry Etching of GaN and Related Materials: Comparison of Techniques," *IEEE Journal of Selected Topics of Quantum Electronics*, Vol. 4, No. 3, 1998, pp. 557–563.

[8] Smith, S. A., et al., "High Rate and Selective Etching of GaN, AlGaN, and AlN Using an Inductively Coupled Plasma," *Applied Physics Letters*, Vol. 71, No. 25, 1997, pp. 3631–3633.

[9] Ohta, H., et al., "Ion-Irradiation Damage on GaN p-n Junction Diodes by Inductively Coupled Plasma Etching and Its Recovery by Thermal Treatment," *Nuclear Instruments and Methods in Physics Research B*, Vol. 409, 2017, pp. 65–68.

[10] Cao, L. H., B. H. Li, and J. H. Zhao, "Etching of SiC Using Inductively Coupled Plasma," *Journal of the Electrochemical Society*, Vol. 145, No. 10, 1998, pp. 3609–3612.

[11] Wang, J. J., et al., "Inductively Coupled Plasma Etching of Bulk 6H-SiC and Thin-Film SiCN in NF_3 Chemistries," *Journal of Vacuum Science and Technology A*, Vol. 16, No. 4, 1998, pp. 2204–2209.

[12] Khan, F. A., and I. Adesida, "High Rate Etching of SiC Using Inductively Coupled Plasma Reactive Ion Etching in SF_6-Based Gas Mixtures," *Applied Physics Letters*, Vol. 75, No. 15, 1999, pp. 2268–2270.

[13] Jiang, L. D., et al., "Inductively Coupled Plasma Etching of SiC in SF_6/O_2 and Etch-Induced Surface Chemical Bonding Modifications," *Journal of Applied Physics*, Vol. 93, No. 3, 2003, pp. 1376–1383.

[14] Mikami, H., et al., "Role of Hydrogen in Dry Etching of Silicon Carbide Using Inductively Coupled and Capacitively Coupled Plasma," *Japanese Journal of Applied Physics*, Vol. 44, No. 6A, 2005, pp. 3817–3821.

[15] Kosugi, R., et al., "Strong Impact of Slight Trench Direction Misalignment from [$11\bar{2}0$] on Deep Trench Filling Epitaxy for SiC Super-Junction Devices," *Japanese Journal of Applied Physics*, Vol. 56, 2017, pp. 04CR05-1–04CR05-4.

[16] Zhuang, D., and J. H. Edgar, "Wet Etching of GaN, AlN, and SiC," *Materials Science and Engineering Reports*, Vol. 48, 2005, pp. 1–46.

[17] Palacios, T., et al., "Wet Etching of GaN Grown by Molecular Beam Epitaxy on Si(111)," *Semiconductor Science and Technology*, Vol. 15, No. 10, 2000, pp. 996–1000.

[18] Rouviere, J. L, et al., "Polarity Determination for GaN Films Grown on (0001) Sapphire and High-Pressure-Grown GaN Single Crystals," *Applied Physics Letters*, Vol. 73, No. 6, 1998, pp. 668–670.

[19] Li, D., et al., "Selective Etching of GaN Polar Surface in Potassium Hydroxide Solution Studied by X-Ray Photoelectron Spectroscopy," *Journal of Applied Physics*, Vol. 90, No. 8, 2001, pp. 4219–4223.

[20] Morimoto, Y., "Few Characteristics of Epitaxial GaN–Etching and Thermal Decomposition," *Journal of the Electrochemical Society*, Vol. 121, No. 10, 1974, pp. 1383–1384.

[21] Shintani, A. and S. Minagawa, "Etching of GaN Using Phosphoric Acid," *Journal of the Electrochemical Society*, Vol. 123, No. 5, 1976, pp. 706–713.

[22] Kim, B. J., et al., "Wet Etching of (0001) GaN/Al_2O_3 Grown by MOVPE," *Journal of Electronic Materials*, Vol. 27, No. 5, 1998, pp. L32–L34.

[23] Hong, S. K., et al., "Evaluation of Nanopipes in MOCVD Grown (0001) GaN/Al_2O_3 by Wet Chemical Etching," *Journal of Crystal Growth*, Vol. 191, No. 1–2, 1998, pp. 275–278.

[24] Mynbaeva, M. G., et al., "Wet Chemical Etching of GaN in H_3PO_4 with Al ions," *Electrochemical and Solid-State Letters*, Vol. 2, No. 8, 1999, pp. 404–406.

[25] Katsuno, M., et al., "Mechanism of Molten KOH Etching of SiC Single Crystals: Comparative Study with Thermal Oxidation," *Japanese Journal of Applied Physics*, Vol. 38, No. 8, 1999, pp. 4661–4665.

[26] Stagg, J. P., "Drift Mobilities of Na^+ and K^+ Ions in SiO_2 Films," *Applied Physics Letters*, Vol. 31, No. 8, 1977, pp. 532–534.

[27] Kodama, M., "GaN-Based Trench Gate Metal Oxide Semiconductor Field-Effect Transistor Fabricated with Novel Wet Etching," *Applied Physics Express*, Vol. 1, 2008, pp. 021104-1–021104-3.

[28] Weyher, J. L., et al., "Recent Advances in Defect-Selective Etching of GaN," *Journal of Crystal Growth*, Vol. 210, No. 1–3, 2000, pp. 151–156.

[29] Minsky, M. S., M. White, and E. L. Hu, "Room-Temperature Photoenhanced Wet Etching of GaN," *Applied Physics Letters*, Vol. 68, No. 11, 1996, pp. 1531–1533.

[30] Lee, J.-S., et al., "Photoelectrochemical Gate Recess Etching for the Fabrication of AlGaN/GaN Heterostructure Field Effect Transistor," *Japanese Journal of Applied Physics*, Vol. 40, No. 3A, 2001, pp. L198–L200.

[31] Chang, W. H., "Micromachining of p-Type 6H-SiC by Electrochemical Etching," *Sensors and Actuators A*, Vol. 112, No. 1, 2004, pp. 36–43.

[32] Ke, Y., et al., "Surface Polishing by Electrochemical Etching of p-Type 4H-SiC," *Journal of Applied Physics*, Vol. 106, No. 6, 2009, pp. 064901-1–064901-7.

[33] Shor, J. S., et al., "Characterization of Nanocrystallites in Porous p-Type 6H-SiC," *Journal of Applied Physics*, Vol. 76, No. 7, 1994, pp. 4045–4049.

[34] Shor, J. S., R. M. Osgood, and A. D. Kutz, "Photoelectrochemical Conductivity Selective Etch Stops for SiC," *Applied Physics Letters*, Vol. 60, No. 8, 1992, pp. 1001–1003.

[35] Shishkin, Y., W. Choyke, and R. P. Devaty, "Photoelectrochemical Etching of n-Type 4H Silicon Carbide," *Journal of Applied Physics*, Vol. 96, No. 4, 2004, pp. 2311–2322.

[36] Mikami, H., et al., "Analysis of Photoelectrochemical Processes in α-SiC Substrates with Atomically Flat Surfaces," *Japanese Journal of Applied Physics*, Vol. 44, No. 12, 2005, pp. 8329–8332.

[37] Ke, Y., R. P. Devaty, and W. J. Choyke, "Comparative Columnar Porous Etching Studies on n-type 6H-SiC Crystalline Faces," *Physica Status Solidi B*, Vol. 245, No. 7, 2008, pp. 1396–1403.

[38] Kasai, H., et al., "Nitrogen Ion Implantation Isolation Technology for Normally-Off GaN MISFETs on p-GaN Substrate," *Physica Status Solidi C*, Vol. 11, No. 3-4, 2014, pp. 914–917.

[39] Negoro, Y., et al., "Electronic Behaviors of High-Dose Phosphorus-Ion Implanted 4H-SiC (0001)," *Journal of Applied Physics*, Vol. 96, No. 1, 2004, pp. 224–228.

[40] Nomoto, K., et al., "Ion Implantation into GaN and Implanted GaN Power Transistors," *ECS Transactions*, Vol. 69, No. 11, 2015, pp. 105–112.

[41] Nomoto, K., et al., "Remarkable Reduction of On-Resistance by Ion Implantation in GaN/AlGaN/GaN HEMTs with Low Gate Leakage Current," *IEEE Electron Device Letters*, Vol. 28, No. 11, 2007, pp. 939–941.

[42] Taguchi, S., et al., "High Threshold Voltage Normally-Off GaN MISFETs Using Self-Aligned Technique," *Physica Status Solidi C*, Vol. 9, No. 3-4, 2012, pp. 858–860.

[43] Ottaviani, L., et al., "Aluminum Multiple Implantations in 6H-SiC at 300K," *Solid-State Electronics*, Vol. 43, No. 12, 1999, pp. 2215–2223.

[44] Park, C., et al., "Paradoxical Boron Profile Broadening Caused by Implantation Through a Screen Oxide Layer," *International Electron Devices Meeting*, Washington, D.C., 1991, pp. 67–70.

[45] Morris, S. J., et al., "An Accurate and Efficient Model for Boron Implants Through Thin Oxide Layers into Single-Crystal Silicon," *IEEE Transactions on Semiconductor Manufacturing*, Vol. 8, No. 4, 1995, pp. 408–413.

[46] Kuroda, N., et al., "Step-Controlled VPE Growth of SiC Single Crystals at Low Temperatures," *Solid State Devices and Materials*, Tokyo, 1987, pp. 227–230.

[47] Mochizuki, K., et al., "Detailed Analysis and Precise Modeling of Multiple-Energy Al Implantations Through SiO_2 Layers into 4H-SiC," *IEEE Transactions on Electron Devices*, Vol. 55, No. 8, 2008, pp. 1997–2003.

[48] Pearson, K., "Contributions to the Mathematical Theory of Evolution, II: Skew Variation in Homogeneous Material," *Philosophical Transactions of the Royal Society of London, A*, Vol. 186, 1895, pp. 343–414.

[49] Janson, M. S., et al., "Ion Implantation Range Distributions in Silicon Carbide," *Journal of Applied Physics*, Vol. 93, No. 11, 2003, pp. 8903–8909.

[50] Stief, R., et al., "Range Studies of Aluminum, Boron, and Nitrogen Implants in 4H-SiC," *International Conference on Ion Implantation Technology*, Kyoto, June 1998, pp. 760–763.

[51] Lee, S.-S. and S.-G. Park, "Empirical Depth Profile Model for Ion Implantation in 4H-SiC," *Journal of Korean Physical Society*, Vol. 41, No. 5, 2002, pp. L591–L593.

[52] Tasch, A. F., et al., "An Improved Approach to Accurately Model Shallow B and BF_2 Implants in Silicon," *Journal of the Journal of Electrochemical Society*, Vol. 136, No. 3, 1989, pp. 810–814.

[53] Suzuki, K., et al., "Comprehensive Analytical Expression for Dose Dependent Ion-Implanted Impurity Concentration Profiles," *Solid-State Electronics*, Vol. 42, No. 9, 1998, pp. 1671–1678.

[54] Mochizuki, K. and H. Onose, "Dual-Pearson Approach to Model Ion-Implanted Al Concentration Profiles for High-Precision Design of High-Voltage 4H-SiC Power Devices," *Materials Science Forum*, Vol. 600–603, 2009, pp. 607–610.

[55] Mochizuki, K., and N. Yokoyama, "Two-Dimensional Modeling of Aluminum-Ion Implantation into 4H-SiC," *Materials Science Forum*, Vol. 679–680, 2011, pp. 405–408.

[56] Mochizuki, K., and N. Yokoyama, "Two-Dimensional Analytical Model for Concentration Profiles of Aluminum Implanted into 4H-SiC (0001)," *IEEE Transactions on Electron Devices*, Vol. 58, 2011, pp. 455–459.

[57] http://www.silvaco.com/products/tcad/process_simulation/process_simulation.html.

[58] https://www.synopsys.com/silicon/tcad/process-simulation.html.

[59] Mochizuki, K., et al., "A-Commercial-Simulator-Based Numerical-Analysis Methodology for 4H-SiC Power Device Formed on Misoriented (0001) Substrates," *IEEE Journal of the Electron Devices Society*, Vol. 3, 2015, pp. 316–322.

[60] Ahmed, S., et al., "Empirical Depth Simulator for Ion Implantation in 6H-SiC," *Journal of Applied Physics*, Vol. 77, No. 12, 1995, pp. 6194–6200.

[61] Rao, M. V., et al., "Donor Ion-Implantation Doping into SiC," *Journal of Electronic Materials*, Vol. 28, No. 3, 1999, pp. 334–340.

[62] Janson, M. S., et al., "Range Distributions of Implanted Ions in Silicon Carbide," *Materials Science Forum*, Vol. 389–393, 2002, pp. 779–782.

[63] Kimoto, T., and J. A. Cooper, *Fundamentals of Silicon Carbide Technology*, Singapore: John Wiley & Sons, 2014, pp. 191–193.

[64] Wilson, R. G., et al., "Redistribution and Activation of Implanted S, Se, Te, Be, Mg, and C in GaN," *Journal of Vacuum Science and Technology A*, Vol. 17, No. 4, 1999, pp. 1226–1229.

[65] Kosugi, R., et al., "First Experimental Demonstration of SiC Super-Junction (SJ) Structure by Multiepitaxial Method," *International Symposium on Power Semiconductor Devices and ICs*, Waikoloa, June 2014, pp. 346–349.

[66] Okamoto, D., et al., "Improved Channel Mobility in 4H-SiC MOSFETs by Boron Passivation," *IEEE Electron Device Letters*, Vol. 35, No. 12, 2014, pp. 1176–1178.

[67] Mokhov, E. N., E. E. Goncharov, and G. G. Ryabova, "Diffusion of Boron in p-Type Silicon Carbide," *Soviet Physics–Semiconductors*, Vol. 18, 1984, pp. 27–30.

[68] Bracht, H., N. A. Stolwijl, and G. Pensl, "Diffusion of Boron in Silicon Carbide: Evidence for the Kick-Out Mechanism," *Applied Physics Letters*, Vol. 77, No. 20, 2000, pp. 3188–3120.

[69] Rüschenschmidt, K., et al., "Self-Diffusion in Isotopically Enriched Silicon Carbide and Its Correlation with Dopant Diffusion," *Journal of Applied Physics*, Vol. 96, No. 3, 2004, pp. 1458–1463.

[70] Rurali, R., et al., "Theoretical Evidence for the Kick-Out Mechanism for B Diffusion in SiC," *Applied Physics Letters*, Vol. 81, No. 16, 2002, pp. 2989–2991.

[71] Bockstedte, M., A. Mattausch, and O. Pankratov, "Ab Initio Study of the Migration of Intrinsic Defects in 3C-SiC," *Physical Review B*, Vol. 68, No. 20, 2003, pp. 205201-1–205201-17.

[72] Gao, H., et al., "Atomistic Study of Intrinsic Defect Migration in 3C-SiC," *Physical Review B*, Vol. 69, No. 24, 2004, pp. 245205-1–245205-5.

[73] Duijin-Arnold, A. V., et al., "Electronic Structure of the Deep Boron Acceptor in Boron-Doped 6H-SiC," *Physical Review B*, Vol. 57, No. 3, 1998, pp. 1607–1619.

[74] Duijin-Arnold, A. V., et al., "Spatial Distribution of the Electronic Wave Function of the Shallow Boron Acceptor in 4H- and 6H-SiC," *Physical Review B*, Vol. 60, No. 23, 1999, pp. 15829–15847.

[75] Aradi, B., et al., "Boron Centers in 4H-SiC," *Materials Science Forum*, Vol. 353–356, 2001, pp. 455–458.

[76] Bockstedte, M., A. Mattausch, and O. Pankratov, "Boron in SiC: Structure and Kinetics," *Materials Science Forum*, Vol. 353–356, 2001, pp. 447–450.

[77] Srindhara, S. G., et al., "Photoluminescence and Transport Studies of Boron in 4H-SiC," *Journal of Applied Physics*, Vol. 83, No. 12, 1998, pp. 7909–7920.

[78] Sadigh, B., et al., "Mechanism of Boron Diffusion in Silicon: An Ab Initio and Kinetic Monte Carlo Study," *Physical Review Letters*, Vol. 83, No. 21, 1999, pp. 4341–4344.

[79] Windle, W., et al., "First Principles Study of Boron Diffusion in Silicon," *Physical Review Letters*, Vol. 83, No. 21, 1999, pp. 4345–4348.

[80] Linnarsson, M. K., et al., "Aluminum and Boron Diffusion in 4H-SiC," *Materials Research Society Proceedings*, Vol. 742, Warrendale, Dec. 2002, paper K6.1.

[81] Linnarsson, M. K., et al., "Boron Diffusion in Intrinsic, n-Type amd p-Type 4H-SiC," *Materials Science Forum*, Vol. 457–460, 2004, pp. 917–920.

[82] Mochizuki, K., H. Shimizu, and N. Yokoyama, "Dual-Sublattice Modeling and Semi-Atomistic Simulation of Boron Diffusion in 4H-Silicon Carbide," *Japanese Journal of Applied Physics*, Vol. 48, 2009, pp. 031205-1–031205-6.

[83] Bockstedte, M., A. Mattausch, and O. Pankratov, "Different Roles of Carbon and Silicon Interstitials in the Interstitial-Mediated Boron Diffusion in SiC," *Physical Review B*, Vol. 70, No. 11, 2004, pp. 115203-1–115203-13.

[84] Bracht, H., "Self- and Foreign-Atom Diffusion in Semiconductor Isotope Heterostructures. I. Continuum Theoretical Calculations," *Physical Review B*, Vol. 75, No. 3, 2007, pp. 035210-1–035210-16.

[85] Janson, M. S., et al., "Transient Enhanced Diffusion of Implanted Boron in 4H-Silicon Carbide," *Applied Physics Letters*, Vol. 76, No. 11, 2000, pp. 1434–1436.

[86] Plummer, G. H., M. D. Deal, and P. B. Griffin, *Silicon VLSI Technology*, Upper Saddle River, NJ: Prentice Hall, 2000, p. 411.

[87] Baliga, B. J., *Silicon Carbide Power Devices*, Singapore: World Scientific, 2005, pp. 978–981.

[88] Lee, C.-T., H.-Y. Lee, and H.-W. Chen, "GaN MOS Device Using SiO_2–Ga_2O_3 Insulator Grown by Photoelectrochemical Oxidation Method," *IEEE Electron Device Letters*, Vol. 24, No. 2, 2003, pp. 54–56.

[89] Huang, L.-S., et al., "AlGaN/GaN metal-oxide-semiconductor High-Electron Mobility Transistors Using Oxide Insulator Grown by Photoelectrochemical Oxidation Method," *IEEE Electron Device Letters*, Vol. 29, No. 4, 2008, pp. 284–286.

[90] Kim, H., S.-J. Park, and H. Hwang, "Thermally Oxidized GaN Film for Use as Gate Insulators," *Journal of Vacuum Science and Technology B*, Vol. 19, No. 2, 2001, pp. 579–581.

[91] Wei, W., et al., "Valence Band Offset of β-Ga_2O_3/Wurtzite GaN Heterostructure Measured by X-Ray Photoelectron Spectroscopy," *Nanoscale Research Letters*, Vol. 7, 2012, pp. 562-1–562-5.

[92] Hua, M., et al., "Integration of LPCVD-SiN_x Gate Dielectric with Recessed-Gate E-Mode GaN MIS-FETs: Toward High Performance, High Stability and Long TDDB Lifetime," *International Electron Devices Meeting*, San Francisco, Dec. 2016, pp. 260–263.

[93] Deal, B. E., and A. S. Grove, "General Relationship for the Thermal Oxidation of Silicon," *Journal of Applied Physics*, Vol. 36, No. 12, 1965, pp. 3770–3778.

[94] Song, Y., et al., "Modified Deal–Grove Model for the Thermal Oxidation of Silicon Carbide," *Journal of Applied Physics*, Vol. 95, No. 9, 2004, pp. 4953–4957.

[95] Hijikata, Y., Y. Yaguchi, and S. Yoshida, "A Kinetic Model of Silicon Carbide Oxidation Based on the Interface Silicon and Carbon Emission Phenomenon," *Applied Physics Express*, Vol. 2, 2009, pp. 021203-1–021203-3.

[96] Watanabe, H., et al., "Energy Band Structure of SiO_2/4H-SiC Interfaces and Its Modulation Induced by Intrinsic and Extrinsic Interface Transfer," *Materials Science Forum*, Vol. 679–680, 2011, pp. 386–389.

[97] Noborio, M, et al., "Reliability of Nitrided Gate Oxide for n- and p-Type 4H-SiC(0001) Metal-Oxide-Semiconductor Devices," *Japanese Journal of Applied Physics*, Vol. 50, No. 9R, 2011, pp. 090201-1–090201-3.

[98] Kimoto, T., et al., "Interface Properties of Metal-Oxide-Semiconductor Structures on 4H-SiC{0001} and $(11\bar{2}0)$ Formed by N_2O Oxidation," *Japanese Journal of Applied Physics*, Vol. 44, No. 3, 2005, pp. 1213–1218.

[99] Fan, Z., et al., "Very Low Resistance Multilayer Ohmic Contact to n-GaN," *Applied Physics Letters*, Vol. 68, No. 12, 1996, pp. 1672–1674.

[100] Liu, Q. Z., and S. S. Lau, "A Review of the Metal–GaN Contact Technology," *Solid-State Electronics*, Vol. 42, No. 5, 1998, pp. 677–691.

[101] Cole, M. W, et al., "Post Growth Rapid Thermal Annealing of GaN: The Relationship Between Annealing Temperature, GaN Crystal Quality, and Contact–GaN Interfacial Structure," *Applied Physics Letters*, Vol. 71, No. 20, 1997, pp. 3004–3006.

[102] Lee, J.-L., et al., "Ohmic Contact Formation Mechanism of Nonalloyed Pd Contacts to p-Type GaN Observed by Positron Annihilation Spectroscopy," *Applied Physics Letters*, Vol. 74, No. 16, 1999, pp. 2289–2291.

[103] Hasegawa, H., and H. Ohno, "Unified Disorder Induced Gap State Model for Insulator-Semiconductor and Metal-Semiconductor Interfaces," *Journal of Vacuum Science and Technology B*, Vol. 4, No. 4, 1986, pp. 1130–1138.

[104] Kim, D.-W., et al., "Electrical Properties of Pd-Based Ohmic Contact to p-GaN," *Journal of Vacuum Science and Technology B*, Vol. 19, No. 3, 2001, pp. 609–614.

[105] Adivarahan, V., et al., "Very-Low-Specific-Resistance Pd/Ag/Au/Ti/Au Alloyed Ohmic Contact to p GaN for High-Current Devices," *Applied Physics Letters*, Vol. 78, No. 18, 2001, pp. 2781–2783.

[106] Porter, L. M., and R. F. Davis, "A Critical Review of Ohmic and Rectifying Contacts for Silicon Carbide," *Materials Science and Engineering B*, Vol. 34, No. 2–3, 1995, pp. 83–105.

[107] Kimoto, T., and J. A. Cooper, *Fundamentals of Silicon Carbide Technology*, Singapore: John Wiley & Sons, 2014, pp. 259–261.

[108] Reshanov, S. A., et al., "Effect of an Intermediate Graphite Layer on the Electronic Properties of Metal/SiC Contacts," *Physica Status Solidi B*, Vol. 245, No. 7, 2008, pp. 1369–1377.

[109] Crofton, J., et al., "Titanium and Aluminum-Titanium Ohmic Contacts to p-Type SiC," *Solid-State Electronics*, Vol. 41, No. 11, 1997, pp. 1725–1729.

[110] Vassilevski, K., et al., "Phase Formation at Rapid Thermal Annealing of Al/Ti/Ni Ohmic Contacts on 4H-SiC," *Materials Science and Engineering B*, Vol. 80, No. 1–3, 1997, pp. 370–373.

[111] Crofton, J., et al., "Contact Resistance Measurement on p-Type 6H-SiC," *Applied Physics Letters*, Vol. 62, No. 4, 1997, pp. 384–386.

[112] Kassamokova, L., et al., "Thermostable Ohmic Contacts on p-Type SiC," *Materials Science Forum*, Vol. 264–268, 1998, pp. 787–790.

[113] Fursin, L. G, J. H. Zhao, and M. Weiner, "Nickel Ohmic Contacts to p- and n-Type 4H-SiC," *Electronics Letters*, Vol. 37, No. 17, 2001, pp. 1092–1093.

[114] Lee, S. K., C. M. Zettering, and M. Östling, "Low Resistivity Ohmic Titanium Carbide Contacts to n- and p-Type 4H-Silicon Carbide," *Solid-State Electronics*, Vol. 44, No. 7, 2000, pp. 1179–1186.

[115] Ohno, H., et al., "Chemical Vapor Deposition of CeO_2 Films Using a Liquid Metalorganic Source," *Electrochemical and Solid-State Letters*, Vol. 9, No. 3, 2006, pp. G87–G89.

[116] Tagui, K., et al., "The Electrical Property of CeO_2 Films Deposited by MOCVD on Si (100)," *Electrochemical and Solid-State Letters*, Vol. 10, No. 7, 2007, pp. D73–D75.

[117] Matsumura, T., et al., "MOCVD of CeO$_2$ and SiO$_2$ Mixture Films Using Alkoxy Sources," *ECS Solid State Letters,* Vol. 4, No. 12, 2015, pp. N17–N19.

[118] Yoshino, M., et al., "High-*k* Dielectric Passivation for GaN Diode with a Field Plate Termination," *Electronics,* Vol. 5, No. 2, 2016, pp. 15-1–15-7.

[119] Song, Q.-W., et al., "Simulation Study on 4H-SiC Power Devices with High-*k* Dielectric FP Terminations," *Diamond & Related Materials,* Vol. 22, 2012, pp. 42–47.

第 8 章
金属半导体接触和单极功率二极管

8.1 引言

如第 1 章所述,功率二极管是电力电子线路最基本的组成单元之一,加有正向电压 V_F 时,电流 I_F 呈现正向特性(见图 1.8)。就双极二极管而言,qV_F 值可以近似等于半导体的带隙能量 E_g,其中 q 是基本电荷($1.6×10^{-19}$C)。例如,可以从以线性比例绘制的图 8.1b 中了解到所测得 GaN p⁺n 二极管(见图 3.6)的正向电流/电压特性(见图 8.1a)[1]。当 I_F 为 5.2mA(即 180A/cm²)时,V_F 为 GaN 的 E_g/q(即 3.44V[2])。因此,为了进一步降低 V_F,必须使用单极二极管。单极二极管的另一个优点是由于没有少数载流子的注入,开关速度非常快而且开关损耗很低。

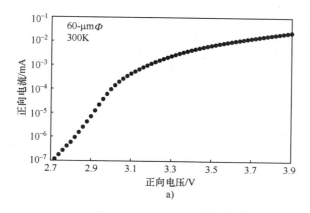

图 8.1 室温下 GaN p⁺n 二极管的正向电流电压特性
a)半对数坐标[1] 和 b)线性坐标

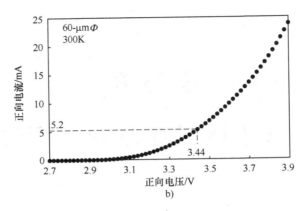

图 8.1 室温下 GaN p⁺n 二极管的正向电流电压特性
a) 半对数坐标[1] 和 b) 线性坐标（续）

单极二极管采用了被称为肖特基结的金属半导体接触[3]。图 8.2a 显示了未接触的金属和 n 型半导体能带图。真空或自由电子的能量可以作为参考能级，代表电子不受金属或半导体影响时所具有的能量。真空能级和费米能级之间的差值（请参见 2.3 节）称为功函数，通常用 $q\Phi$ 来表示，以 eV 为单位（或用 Φ 表示，以 V 为单位）。就半导体而言，功函数就是掺杂浓度的函数。由于电子亲和能 qX（即真空能级与导带底能量 E_C 之差）是恒定不变的，因此金属和半导体的功函数分别表示为 $q\Phi_m$ 和 $q(X+\Phi_n)$，如图 8.2a 所示，其中 $q\Phi_n$ 为半导体 E_C 和费米能级 E_F 之差。此外，由于 $\Phi_m > X + \Phi_n$，因此半导体中电子的平均总能量比金属中电子的能量更高。

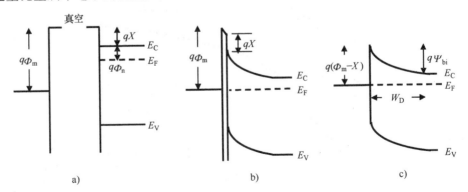

图 8.2 金属和 n 型半导体能带图
a) 接触前，b) 紧密接触，c) 无间隙

当金属和半导体紧密接触并以有限的间隙分开时（见图 8.2b），金属的费米能级等于 $q\Phi_m$，与处于平衡状态的半导体的 E_F 一致。请注意，图 8.2b 中显示，真空能级是连续的，电子亲和能是恒定不变的。如果金属和半导体之间的间隙变为零（见图 8.2c），则在金属-半导体界面处会出现突变的 E_C 不连续性。这种不连续性的程度被称为肖特基势垒高度（$q\Phi_B$），也可表示为

$$q\Phi_B = q(\Phi_m - X) = q(\Psi_{bi} + \Phi_n) \tag{8.1}$$

由于 $q\Phi_B$ 小于半导体的 $E_g(=E_C-E_V)$ 值，因此，只要通过漂移层的电压降可忽略不计，单极二极管的 V_F 就会小于双极二极管的 V_F。

8.2 肖特基势垒的降低

肖特基势垒的降低是指在电场存在下由镜像力引起的 $q\Phi_B$ 的降低。具体来说，当电子与金属表面的距离为 x 时，金属表面上将感应出正电荷。电子与正电荷之间的吸引力等于电子与位于 $-x$ 处的等价正电荷（即镜像电荷）之间的吸引力 F，可表示为

$$F = -q^2/[4\pi\varepsilon_r\varepsilon_o(2x)^2] = -q^2/(16\pi\varepsilon_r\varepsilon_o x^2) \tag{8.2}$$

电子在从点 x 转移到无穷大的过程中所做的功为

$$W(x) = \int_x^\infty F dx = -q^2/(16\pi\varepsilon_r\varepsilon_o x^2) \tag{8.3}$$

式（8.3）代表距金属表面 x 处的电子的势能 $E_p(x)$。

在电场 E 的作用下，$E_p(x)$ 从 $-q|E|x$ 开始降低，如下所示

$$E_p(x) = -q|E|x - q^2/(16\pi\varepsilon_r\varepsilon_o x) \tag{8.4}$$

$E_p(x)$ 在 x_m 处达到最大值，x_m 如下所示

$$x_m = [q/(16\pi\varepsilon_r\varepsilon_o|E|)]^{0.5} \tag{8.5}$$

当 $dE_p/dx = 0$ 时，$q\Phi_B$ 从 $x=0$ 的值［见式（8.1）］下降了 $\Delta q\Phi_B$，可以表示为

$$\Delta q\Phi_B = (qE_{max}/4\pi\varepsilon_r\varepsilon_o)^{0.5} \tag{8.6}$$

式中，E_{max} 为最大电场强度（见 3.5 节）。

8.3 正向偏置的肖特基结

肖特基结中的正向电流存在两种截然不同的机制：肖特基势垒的载流子热离子发射理论[4]和半导体到金属的载流子扩散理论[3]（见 8.4 节）。热离子发

射理论假设只有电子能量大于或等于金属-半导体界面处的导带能量,才会形成式(8.7)中的正向电流 J_{TE}。另一方面,扩散理论假设式(8.8)中的正向电流 J_D 是由耗尽层的长度驱动的。

$$J_{TE} = qN_C(kT/2\pi m_n)^{0.5}\exp(-q\Phi_B/kT)[\exp(qV_F/kT)-1] \quad (8.7)$$

和

$$J_D = (q^2D_nN_C/kT)\{[2q(\Phi_B-\Phi_n-V_F)N_D]/\varepsilon_r\varepsilon_o\}^{0.5}$$
$$\times \exp(-q\Phi_B/kT)[\exp(qV_F/kT)-1] \quad (8.8)$$

式中,N_C 为导带中的有效态密度 [见式(2.6b)];m_n 为电子的有效质量;D_n 为电子的扩散率(见3.2节)。

对于硅肖特基结,正向电流受 J_{TE} 限制。但是 GaN 和 4H-SiC 肖特基结不是如此,原因如下:金属-半导体结的电荷、场分布以及能带图类似于图 3.4 中所示的 p$^+$n 结的情况。当金属和半导体之间加正向压降 V_F 并且半导体接地时,通过式(3.28)和式(3.31),E_{max} 可表示为

$$E_{max} = [2qN_D(\Psi_{bi}-V_F)/\varepsilon_r\varepsilon_o]^{0.5} \quad (8.9)$$

例如,当击穿电压 BV 约为 0.5kV 时,硅肖特基结的 N_D 为 $5\times10^{14}\text{cm}^{-3}$;GaN 和 4H-SiC 肖特基结的 N_D 为 $1\times10^{17}\text{cm}^{-3}$[5]。当 $\Psi_{bi}-V_F$ 为 0.1V 时,硅肖特基结的 E_{max} 为 4kV/cm,GaN 和 4H-SiC 肖特基结的 E_{max} 为 60kV/cm。虽然硅肖特基结的电子漂移速度 (v_n) 几乎与 $|E|$ 成正比,但 GaN 和 4H-SiC 肖特基结的 v_n 几乎饱和(见图 8.3)[6-8]。

在较高的 E_{max} 下,爱因斯坦关系 [见式(3.6a)]变化很小,如下所示

$$D_n = (akT/q)\mu_n(E) \quad (8.10)$$

其中,对于 n 型 GaN,在 E_{max} 为 60kV/cm 的情况下,$a = 1.5$[9]。式(8.8)变为

$$J_D = aq\mu_n(E_{max})N_CE_{max}\exp(-q\Phi_B/kT)[\exp(qV_F/kT)-1] \quad (8.11)$$

对于 n 型 GaN 和 4H-SiC,当 E_{max} 为 60kV/cm 时,μ_n 分别为 345cm^2/Vs 和 164cm^2/Vs(见图 8.3)。这些高场 μ_n 值远低于低场 μ_n^{GaN}(1000cm^2/Vs)和 μ_n^{4H-SiC}(1140cm^2/Vs)的值[10]。这种情况意味着电子与晶格的碰撞(即扩散)可有效地帮助 n 型 GaN 或 4H-SiC 中的电子到达肖特基结(见图 8.4)[11-13]。然而,在 GaN 和 SiC 功率器件的文献中,热离子发射被认为是控制一个肖特基结的正向电流传导机制[14-17]。因此,第 8.4 节和第 8.5 节将介绍基于扩散[3]和热离子发射扩散(TED)[18]的正向电流/电压特性。

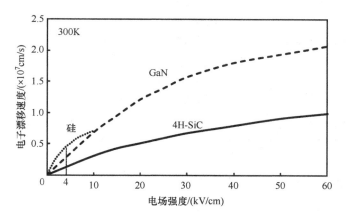

图 8.3 n 型 GaN、4H-SiC 和硅中电子的漂移速度与电场强度的关系[6-8]

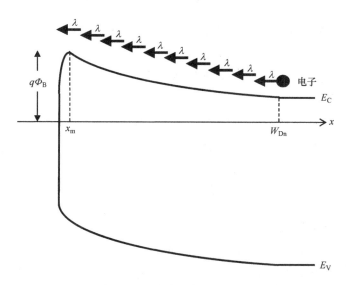

图 8.4 GaN 或 4H-SiC 肖特基结的正向偏置能带示意图
λ 表示电子平均自由程

8.4 基于扩散理论的正向电流/电压特性

根据肖特基[3]提出的扩散理论，假设 W_{Dn} 足够大于 λ（在 GaN 和 4H-SiC 肖特基结中就是这种情况），就可以利用电子的扩散率和迁移率来分析电流/电压特性。根据上述扩散理论，通过积分电子扩散和漂移方程，可以得出肖特基结中正向电流密度（J_D）与正向电压（V_F）之间的关系［见式（3.10a）］，即

$$J_D = qn\mu_n E + qD_n(\partial n/\partial x) \tag{8.12}$$

电流穿过肖特基结附近的耗尽区。假设电势用 Φ 表示，则

$$E = -\partial\Phi/\partial x \tag{8.13}$$

利用爱因斯坦关系式 [见式 (3.6a)]，对于 n 型半导体

$$D_n = (kT/q)\mu_n \tag{8.14}$$

因此，式 (8.12) 又可以写为

$$J_D = qD_n[-(qn/kT)(\partial\Phi/\partial x) + \partial n/\partial x] \tag{8.15}$$

通过在积分之前将式 (8.15) 的两边乘以积分因子 $\exp(-q\Phi/kT)$，并假设 J_D 不是 x 的函数，则式 (8.15) 变为

$$J_D \int_0^{W_{Dn}} \exp(-q\Phi/kT)\mathrm{d}x = qD_n[n\exp(-q\Phi/kT)]_0^{W_{Dn}} \tag{8.16}$$

如果将 $\Phi(0)$ 定义为 0，那么 $\Phi(W_{Dn})$ 则变成 $\Phi_B - \Phi_n - V_F$，利用式 (2.12a) 可分别得到 $n(W_{Dn})$ 和 $n(0)$ 为

$$n(W_{Dn}) = N_C \exp(-q\Phi_n/kT) \tag{8.17a}$$

和

$$n(0) = N_C \exp(-q\Phi_B/kT) \tag{8.17b}$$

将这些边界值代入式 (8.16)，得到

$$J_D = qD_n N_C \exp(-q\Phi_B/kT)[\exp(qV_F/kT) - 1]/\int_0^{W_{Dn}} \exp(-q\Phi/kT)\mathrm{d}x \tag{8.18}$$

通过对式 (3.27b) 积分，可以将 $\Phi(x)$ 近似为

$$\Phi(x) = (qN_D/\varepsilon_r\varepsilon_o)x(W_{Dn} - x/2) \quad (0 < x < W_{Dn}) \tag{8.19}$$

当把式 (8.19) 代入式 (8.18)，并积分替换方程时，J_D 与 V_F 的函数关系可以表示为

$$J_D = J_{DS}[\exp(qV_F/kT) - 1] \tag{8.20a}$$

以及

$$J_{DS} = (q^2 D_n N_C/kT)\{2q(\Phi_B - \Phi_n - V_F)N_D/\varepsilon_r\varepsilon_o\}^{0.5}\exp(-q\Phi_B/kT) \tag{8.20b}$$

在式 (8.20b) 中，随着 V_F 值的改变，J_{DS} 的二次方根值变化很小，因此可以将式 (8.20a) 近似为

$$J_D = J_{DS}'[\exp(qV_F/nkT) - 1] \tag{8.21}$$

式中，J_{DS}' 与 V_F 无关；n 为理想因子 (见 3.6.1.1 节)。

8.5 基于 TED 理论的正向电流/电压特性

8.4 节中显示，即使在电流流动的情况下，肖特基结也被假设为几乎处于热平衡状态。不过在本节中，将通过 x_m 处的热电子复合速度 v_R 这项边界条件来考虑热离子发射对电流/电压特性的影响。这里考虑 $q\varPhi_B$ 足够大以至于金属表面与 $x=W_{Dn}$ 之间的电荷浓度（见图 8.4）主要是电离施主的情况。假设电子在 $x=0$ 到 $x=x_m$ 之间的势垒区域进行吸收，则基于 TED 理论的正向电流密度可以表示为

$$J_{TED} = \{A^*T^2/[1+(v_R/v_D)]\}\exp(-q\varPhi_B/kT)[\exp(qV/kT)-1] \quad (8.22a)$$

$$v_R = (A^*T^2/2q)(2\pi m_n kT/h^2)^{-1.5} \quad (8.22b)$$

$$v_D = (\mu_n kT/q)\exp(-q\varPhi_B/kT) / \int_0^{W_{Dn}} \exp[E_C(x)/kT]dx \quad (8.22c)$$

$$E_C(x) = q\varPhi_B - (q^2 N_D/2\varepsilon_r\varepsilon_o)(2W_{Dn}x - x^2) \quad (8.22d)$$

和

$$A^* = 4\pi m_n k^2 q/h^3 \quad (8.22e)$$

式中，$E_C(x)$ 为从金属费米能级测得的导带边缘能量；v_D 为从 W_{Dn} 到 x_m 与电子传输相关的有效扩散速度；h 为普朗克常数（6.6×10^{-34}Js）；A^* 为有效理查森常数。

对于自由电子，A^* 等于理查森常数 A，为 120Acm^{-2}K^{-2}。然而，由于 GaN 在导带底具有各向同性的有效质量（$0.20\times9.1\times10^{-31}$kg），因此 A^*/A 仅为 0.20。另一方面，对于 4H-SiC 来说，由于其导带极小值处的电子有效质量是 GaN 的 6 倍，A^*/A 约为 1.2[19]。

在式（8.22a）中，v_R/v_D 决定了 TE 与扩散的相对限流因子。在 GaN 肖特基结中，由于 μ_n 随温度的升高而降低[12]，J_{TED} 在高温下受到扩散过程的限制（见 8.7.1 节）。对于采用了离子注入工艺的带屏蔽层的 4H-SiC 肖特基结（见 8.10.2 节），即使在室温下，扩散过程对电流/电压特性的影响仍然很大[13]。请注意，数值分析对于同时考虑 TE 和扩散是很方便的[11]。

8.6 基于 TFE 理论的反向电流/电压特性

当对金属/n 型半导体肖特基结施加负偏压 V_R 时，电压主要施加在 n 型半

导体形成的耗尽区上。GaN 和 4H-SiC 肖特基结的 E_{max} 比硅肖特基结的 E_{max} 大一个数量级,由于有较大的 E_{max},产生的隧穿电流是不可忽略的,费米能级附近的电子隧穿称为场发射(FE),然而对于热激发电子的隧穿,其势垒较 FE 中更窄,因此称为热电子场发射(TFE)(见图 8.5)。当热能 kT 远低于 E_{00} 时,FE 占主导地位,其中 E_{00} 定义为

$$E_{00} = (qh/4\pi)(N_D/m_n\varepsilon_r\varepsilon_o)^{0.5} \qquad (8.23)$$

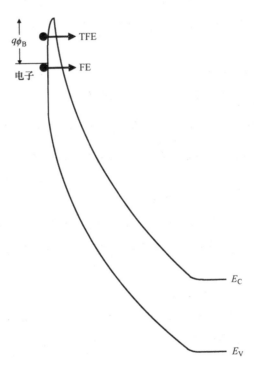

图 8.5 肖特基结反向偏置的电子隧穿(即 FE 和 TFE)能带示意图

由图 8.6 可见,当 $T<15K$ 且 $N_D>10^{15}cm^{-3}$ 时,kT 小于 E_{00},FE 占主导地位。另一方面,在室温下,当 $N_D<10^{16}cm^{-3}$ 时,$kT \gg E_{00}$,TFE 占主导地位。

据先前报道,假设 TFE 占主导,计算出的电流/电压特性与 100μm 直径的 Ni/GaN 肖特基势垒二极管(SBD)的反向特性一致[20],但与 9mm² 尺寸的 Ni/GaN SBD 器件的反向特性不同[21]。此外,对于 4H-SiC SBD,线位错密度会影响所测得的反向特性。例如,对比 53 Mo/4H-SiC SBD 的反向电流/电压特性,其中最大反向电流密度(0.6kV 条件下为 $2 \times 10^{-1}mA/cm^2$)比最小反向电流密度

（0.6kV 条件下为 2×10^{-5}mA/cm²）大四个数量级[22]（见图 8.7）。

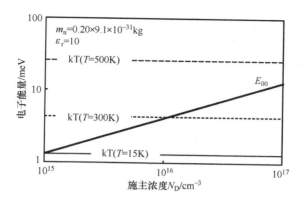

图 8.6　计算出的作为施主浓度函数的 E_{00} 与热电子能量之间的关系

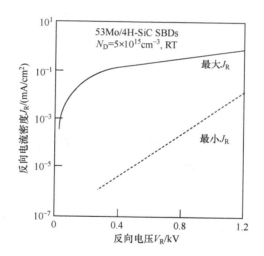

图 8.7　室温下测得的 Mo/4H-SiC SBD 的最大和最小反向电流密度与电压的特性曲线[22]

这些现象表明，在反向偏置条件下描述 J_{TED} 的方程式（8.22a）用处不大。相反，在引入平面 SBD 结构之后，在结构中引入较宽禁带的半导体层或较薄的 p⁺型半导体层有望得到较低的 E（见图 8.8），从而实现反向电流的最小化。

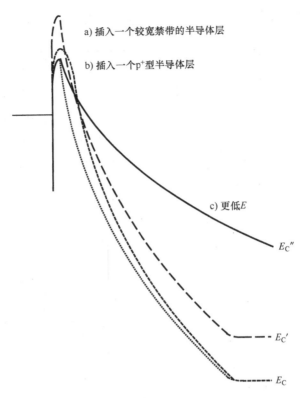

图 8.8 通过反向偏置的肖特基结抑制电子隧穿的导带示意图。在 a) 和 b) 中，在肖特基结处插入了一个禁带宽的半导体层或一个较薄的 p^+ 型层，而在 c) 中，电场（实线）低于图 8.5 所示的电场

8.7 纯 SBD

8.7.1 纯 GaN SBD

2001 年，由于 GaN 衬底的问世使制造 450V 和 700V 垂直型 GaN SBD[23,24]首次成为可能，图 8.9a 显示了纯的 GaN SBD。器件背面欧姆接触是 Ti/Al[23]或 Ti/Al/Pt/Au[24]，肖特基接触是 Pt/Au[23]或 Pt/Ti/Au[24]。但是 N_D 相对较大，分别是大于 $5×10^{16} cm^{-3}$[23]和 $10^{17} cm^{-3}$ 左右[24]。2006 年，采用 N_D 为 $7×10^{15} cm^{-3}$ 的材料首次在 GaN 衬底上制造出 600V 的纯 GaN SBD[25,26]。之后，报道了首款在 n^+ GaN 衬底上制作出具有轻掺杂外延层（$N_D = 1×10^{14} cm^{-3}$）的纯 GaN SBD 器

件,该器件采用了氩离子注入工艺进行边缘隔离(见第11章)[27,28],击穿电压 BV 为1.3kV。

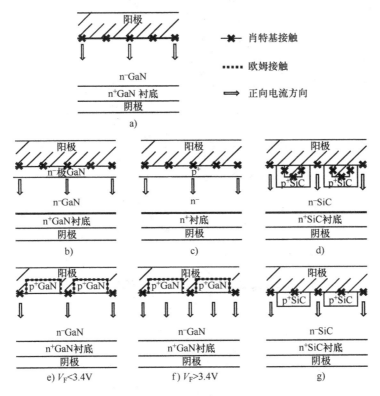

图8.9 SBD有源区横截面示意图:a)纯SBD,b)具有较宽带隙层的SBD,c)具有薄 p^+ 型层的SBD,d)具有较薄沟槽的SBD p^+ 型区域,e、f)GaN混合型pn肖特基(MPS)二极管和g)4H-SiC结势垒肖特基(JBS)二极管

2010年,首次利用高质量的GaN(μ_n = 930cm^2/Vs)制作出纯SBD,采用了Ni/Au作为肖特基接触的金属且采用了场板和边缘终端技术[29](见第11章)。当 BV = 1.1kV, J_F = 500A/cm^2,器件正向电流 I_F 超过10A时, V_F 有相对较低的值1.46V。2015年,制备出同时具有高 I_F (50A)和高 BV (0.79kV)的9mm^2 纯GaN SBD,器件采用了Ni作为肖特基接触金属且采用了台面式场板和边缘终端技术[21](见第11章)。如图8.10a所示,由于半导体封装过程中功耗的典型限制(P_P = 300W/cm^2 [30]), V_F 为1.30V。

随着温度的升高,μ_n 由于声子散射而减小。$q\Phi_B$ 与温度(348~573K)之间的关系可以通过在TED模型的基础上拟合Ni/n-GaN纯SBD的 I_F/V_F 特性进行分析[12]。

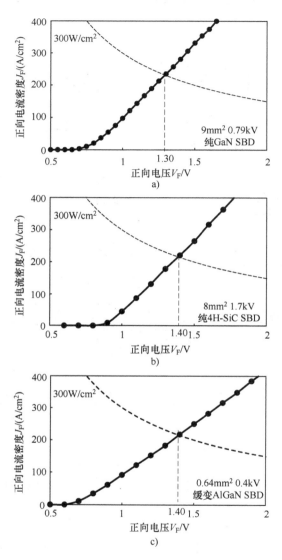

图 8.10 室温下测量的正向电流密度/电压特性
a) 9mm² 0.79kV 纯 GaN SBD[21]，b) 8mm² 1.7kV 纯 4H-SiC SBD[33] 和 c) 缓变 AlGaN SBD[34]

8.7.2 纯 4H-SiC SBD

第一个被报道的具有低 V_F（J_F = 100A/cm² 时为 1.1V）的高压纯 SiC SBD 是多晶 6H SiC 型[31]。其肖特基金属是 Pt，BV 是 400V。BV 后来通过使用 Ar 离

子注入的边缘终端技术得到了改善[32]（参见第 11 章）。对于纯 4H-SiC SBD，当 $J_F=100A/cm^2$，V_F 为 2.4V 时，BV 为 $1.5\sim2.5kV$[35]，当 $J_F=100A/cm^2$，V_F 为 6V 时，BV 为 4kV[36]，当 $J_F=25A/cm^2$，V_F 为 2.4V 时，通过使用 Ni 肖特基接触金属得到了高达 4.9kV 的 BV[37]。此外，还制备了当 $J_F=100A/cm^2$，V_F 为 1.89V 时，BV 为 4.15kV 的纯 Mo/4H-SiC SBD[38]。

以一个具有大 I_F 值的纯 4H-SiC SBD 为例，图 8.10b 显示了 $8mm^2$ 1.7kV SBD 的 J_F-V_F 特性[32]。由图 8.10b 可知，P_P 为 $300W/cm^2$ 时，V_F 为 1.40V。

8.8 缓变 AlGaN SBD

通过采用 Pd[39] 和 Ni/Au[33] 作为肖特基接触金属材料制备了具有缓变 AlGaN 层的 GaN SBD，使垂直型 SBD 具有更宽的带隙层（见图 8.9b）。由于浅施主能级中氧缺陷的存在（如在 $Al_{0.26}Ga_{0.74}N$ 中的能级深度为 0.03eV [40,41]），$q\Phi_B$ 降低［薄层势垒（TSB）模式］[42]，使得分析它们的 J_R/V_R 特征往往还是比较复杂的[39]。如图 8.10c 所示，$0.64mm^2$、0.4kV 缓变 AlGaN SBD[32] 的 V_F 为 1.40V（P_P 为 $300W/cm^2$），比图 8.10a 所示的纯 GaN SBD 的 V_F 高 0.1V。

8.9 带有 p^+ 型薄层的 4H-SiC SBD

对于具有 p^+ 型层（厚度：a）的 SBD（见图 8.9c），净施主浓度（N_D-N_A）的理想分布如图 8.11a 所示。图 8.11b 所示的电场 E 的分布可表示为

$$E=E_{max}-qpx/\varepsilon_r\varepsilon_o, 0<x<a \text{ 时}$$
$$=-qnx(W_{Dn}-x)/\varepsilon_r\varepsilon_o, a<x<W_{Dn}\text{时} \quad (8.24)$$

其中 E_{max} 为

$$E_{max}=q[-pa+n(W_{Dn}-x)]/\varepsilon_r\varepsilon_o \quad (8.25)$$

导带最小值（E_C）在 $x=b$ 处达到最大值，其中

$$b=[ap-(W_{Dn}-x)n]/p \quad (8.26)$$

因此 Φ_B 比无 p^+ 薄层的 SBD 的 Φ_B 增加了

$$\Delta\Phi_B=E_{max}b-qpb^2/2\varepsilon_r\varepsilon_o \quad (8.27)$$

当 $p\gg n$，$ap\gg W_{Dn}n$ 时，$\Delta\Phi_B\approx\Phi_B+qpa^2/2\varepsilon_r\varepsilon_o$；即有效 Φ_B 增加了（见图 8.11c）。

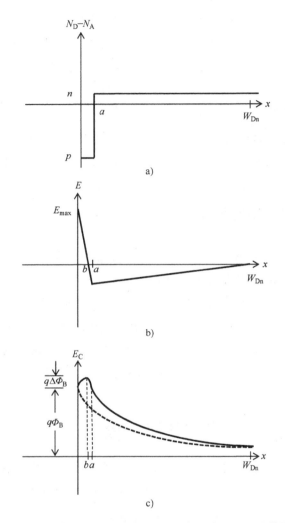

图 8.11 a) 净施主浓度的理想化分布，b) 电场分布和 c) 导带最小值的分布

图 8.9d 证明了具有 p^+ 薄层的 SBD，即 4H-SiC 沟槽 SBD，其中的沟槽已经延伸到 p^+ 层区域。正如在 0.6kV 级 SBD 的情况下所模拟的[43]，能量 $|E|$ 超过 1.5MV/cm 的区域出现在沟槽 SBD 的沟槽底部下方，然而在平面 SBD 中 $|E|$ 出现在肖特基结附近（见图 8.12）。当反向偏压为 0.6kV 时，沟槽 SBD 肖特基结处的 E_{max} 为 0.68MV/cm，比平面 SBD E_{max}（1.65MV/cm）的一半还小[43]。当 n 型漂移层（$N_D = 1.0 \times 10^{16} cm^{-3}$）的厚度为 5μm 时，需要使用 Φ_B 大的金属（例如 1.31V）来抑制 0.6kV 时的反向漏电。由于薄 p^+ 区和沟槽结构的结合，在具有 1.05μm 深沟槽的漂移层上成功地形成了 Φ_B（0.85V）小的金属。也就是说，

V_F 下降了 0.48V,而在 0.6kV 条件下保持相似的反向漏电(即对于 3.06mm² 的芯片,小于 1×10⁻⁶A)[43]。

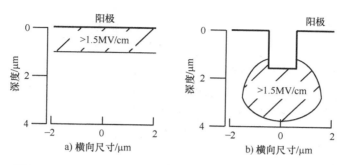

图 8.12 在 0.6kV 反向偏置 a) 平面和 b) 具有 p⁺型薄层的沟槽 SBD 中,电场强度超过 1.5MV/cm 区域的模拟分布[43]

图 8.13 显示了具有 p⁺型区[43]和 0.79kV 纯 GaN SBD[21]的 4H-SiC 沟槽 SBD 的电流/电压特性。在 P_P = 300W/cm² 的条件下,具有 p⁺型区(1.03V)的 4H-SiC 沟槽 SBD 的 V_F 比纯 GaN SBD 的 V_F 低 0.27V(见图 8.13a)。就反向特性而言,两个 SBD 都具有极低的漏电流密度(小于 1mA/cm²)(见图 8.13b);但是,纯 GaN SBD 在 V_R = 0.7kV 时的 J_R 约为带 p⁺型区的 4H-SiC 沟槽 SBD 的 J_R 的六倍。请注意,尽管可以使用 ICP 刻蚀和湿法刻蚀来制造 GaN 沟槽 SBD(见 7.2 节),但是由于难以提高镁受主离子注入的激活率,因此很难制造出具有薄 p⁺型区的 GaN 沟槽 SBD(见 7.3.1 节)。

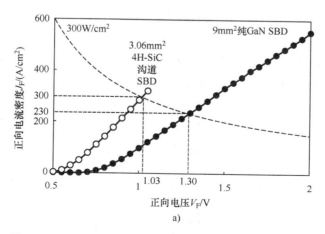

图 8.13 室温下带有 p⁺型薄层[43]和纯 SBD[21]的 0.7kV 级 4H-SiC 沟槽 SBD 的 a) 正向和 b) 反向电流密度/电压特性

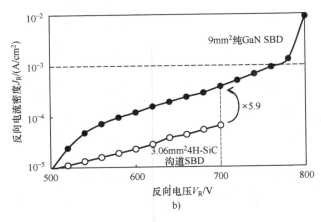

图 8.13 室温下带有 p⁺ 型薄层[43] 和纯 SBD[21] 的 0.7kV 级
4H-SiC 沟槽 SBD 的 a) 正向和 b) 反向电流密度/电压特性（续）

8.10 屏蔽平面 SBD

屏蔽技术可以减小具有薄 p⁺ 型区沟槽 SBD 的肖特基结处的 $|E|$ 值，这种技术也适用于肖特基结和 p-n 结交错的平面 SBD（见图 8.9e~g）。阳极金属和 p-n 结形成欧姆接触的屏蔽平面 SBD 称为混合型 p-n 肖特基（MPS）二极管。另一方面，其阳极金属和 p-n 结形成肖特基接触的屏蔽平面 SBD 称为 JBS 二极管。当 $V_F < E_g/q$ 时，正向电流仅流过 MPS 二极管和 JBS 二极管的肖特基结（见图 8.9e 和 g）。相反，当 $V_F > E_g/q$ 时，正向电流也流过 MPS 二极管的 p-n 结（见图 8.9f）。

众所周知，电子-空穴对复合引起的能量损失常常可以产生缺陷[44]（参见 12.6 节）。由于 4H-SiC 的堆垛层错能量（$0.014J/m^2$[45]）比 GaN（$1.2J/m^2$[46]）小得多（见 2.2.3 节），因此对于 4H-SiC 二极管，JBS 二极管比 MPS 二极管更有优势。另一方面，对于屏蔽 GaN SBD，因为难于在边缘终端区中进行镁离子注入，所以优选有源区中存在 p⁺ 型区的 MPS 二极管。

8.10.1 GaN 混合型 p-n 肖特基二极管

制作了 GaN MPS 二极管（见图 8.9e 和 f），并将它们的电流/电压特性与纯 GaN SBD 和 GaN p-n 结二极管进行了比较[47]。如图 8.14a 所示，GaN MPS 二极管的导通电压（0.8V）与纯 GaN SBD 的导通电压一致。尽管纯 GaN SBD 的 J_F

在 4.4kA/cm² 左右达到饱和，但是由于 p-n 结中有电流流过，GaN MPS 的 J_F 呈指数增长。GaN MPS 二极管的 BV 为 1.6kV，介于纯 GaN SBD 的 BV 和 GaN p-n 结二极管的 BV 之间（见图 8.14b）。

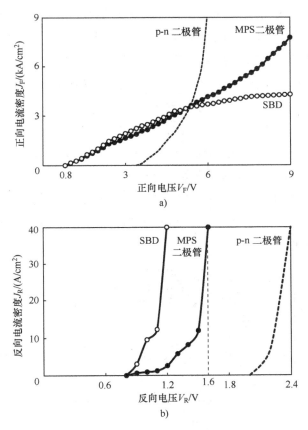

图 8.14 室温下 0.01mm² GaN MPS 二极管的 a) 正向和 b) 反向电流密度/电压特性[47]

8.10.2 4H-SiC JBS 二极管

在 JBS 二极管中，由于 p^+-n 结和肖特基结相互交错，可以屏蔽肖特基结，进而避免受到高电场的影响[48]（见图 8.9g）。

为了阻止导通状态下的单极电导，会选择 p^+ 型区域之间的空间。第一个高压 SiC JBS 二极管于 1998 年发布[49,50]。通过以 $3×10^{15}cm^{-3}$ 掺杂浓度掺杂 9μm 厚的漂移层可实现 0.7kV 的 BV[50]。以 $3×10^{15}cm^{-3}$ 掺杂 27μm 厚的漂移层，4H-SiC JBS 二极管的 BV 可提高到 2.8kV[51]，以 $2×10^{15}cm^{-3}$ 掺杂 30μm 厚的漂移层，其

BV 可提高到 4.3kV[52]，以 $6×10^{14}cm^{-3}$ 掺杂厚度为 120μm 的漂移层，BV 可提高到 10kV[53]。

4H-SiC JBS 二极管已注入铝离子（请参见 7.3.2 节）。因此，通过假设硅 JBS 二极管 p^+-n 结的横向延伸长度与垂直深度的比（r）为 85% 来分析这些 4H-SiC JBS 二极管的 J_F/V_F 特性[54]。对于 4H-SiC JBS 二极管，其比值 r 设定为 1.5%[55]。因为铝的扩散率极低，后一种情况下的 r 值小得多[56]。另一方面，在 4H-SiC 中注入垂直分布的铝的浓度会由于沟道效应而导致尾部延伸（请参见 7.3.2 节）。此外，根据二维蒙特卡洛模拟，注入的铝离子和注入引起的点缺陷从注入掩膜的边缘横向扩散[57-59]。因此，当减小 4H-SiC JBS 二极管中 p^+ 型区之间的间距时，需要考虑到此类影响非常重要。例如，通过实验和计算研究了相距 1μm、具有 2μm 宽的 p^+ 条带区域的 4H-SiC JBS 二极管和镍/4H-SiC SBD[57]。以 $3×10^{15}cm^{-3}$ 浓度掺杂硅的 22μm 厚的漂移层为例，在假设 $q\Phi_B$ 为 1.59eV，μ_n 均匀度为 840cm²/Vs 的模拟下，图 8.15 展现了 1.8mm² 镍/4H-SiC SBD（空心圆）的 J_F/V_F 特性（虚线）。

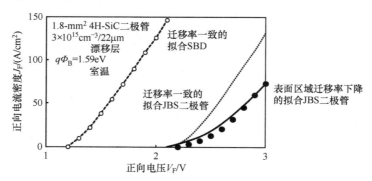

图 8.15 实测（符号）和模拟（线）镍/4H-SiC SBD 和 JBS 二极管的正向电流密度/电压特性，其中以 $3×10^{15}cm^{-3}$ 掺杂了 22μm 厚的漂移层。通过假设厚度为 0.23μm 的表面区域（实线）中 μ_n 一致为 840cm²/Vs（虚线和点线）或 μ_n 降为 450cm²/Vs 时进行仿真

但是，当在相同的假设下模拟测量 1.8mm² 4H-SiC JBS 二极管的 J_F/V_F 特性时，J_F 被高估了（点线）。因此提出了一个两层模型，该模型考虑了注入引起的点缺陷的横向扩展[57]。根据该模型，由于间隙杂质横向扩展明显的区域低于 0.23μm，可以认为表面区域的 μ_n 退化了。通过模拟，成功地复现了 1.8mm² 4H-SiC JBS 二极管的实测 J_F/V_F 特性（实心圆）。该模拟假设在比 0.23μm 浅的区域中 μ_n 为 450cm²/Vs，在大于 0.23μm（虚线）较深区域中 μ_n 为 840cm²/Vs。

8.11 总结

本章的前半部分解释了金属-半导体接触的物理原理，包括肖特基势垒的降低以及基于扩散和热离子发射扩散理论的电流传输特性。与传统的纯 SBD 相比，为了减少反向漏电效应，可以选择具有 p^+ 型薄层、宽带隙层以及具有降低的电场效应的 SBD 结构。本章的后半部分介绍了缓变 AlGaN SBD、具有薄 p^+ 型区的 4H-SiC 沟槽 SBD、GaN MPS 二极管、4H-SiC JBS 二极管以及纯 GaN 和 4H-SiC SBD 功率器件。

参 考 文 献

[1] Mochizuki, K., et al., "Influence of Surface Recombination on Forward Current–Voltage Characteristics of Mesa GaN p+n Diodes Formed on GaN Freestanding Substrates," *IEEE Transactions on Electron Devices*, Vol. 59, No. 4, 2012, pp. 1091–1098.

[2] Monemar, B., et al., "Recombination of Free and Bound Excitons in GaN," *Physica Status Solidi B*, Vol. 245, No. 9, 2008, pp. 1723–1740.

[3] Schottky, W., "Halbleitertheorie der Sperrschicht," *Naturwissenschaften*, Vol. 26, No. 52, 1938, pp. 843–843.

[4] Bethe, H. A., "Theory of the Boundary Layer of Crystal Rectifiers," *MIT Radiation Laboratory Report*, Vol. 185, No. 43, 1943, pp. 12–57.

[5] Baliga, B. J., *Gallium Nitride and Silicon Carbide Power Devices*, Singapore: World Scientific, 2017, p. 79.

[6] Jacoboni, C., et al., "A Review of Some Charge Transport Properties of Silicon," *Solid-State Electronics*, Vol. 20, No. 2, 1977, pp. 77–89.

[7] Bhapkar, U. V., and M. S. Shur, "Monte Carlo Calculation of Velocity-Field Characteristics of Wurtzite GaN," *Journal of Applied Physics*, Vol. 82, No. 4, 1997, pp. 1649–1655.

[8] Khan, I. A, and J. A. Cooper, "Measurement of High Field Electron Transport in Silicon Carbide," *Materials Science Forum*, Vol. 264–268, 1998, pp. 509–512.

[9] Rodrigues, C. G., Á. R. Vasconcellos, and R. Luzzi, "Nonlinear Charge Transport in III-V Semiconductors: Mobility, Diffusion, and a Generalized Einstein Relation," *Journal of Applied Physics*, Vol. 99, 2006, pp. 073701-1–073701-12.

[10] Baliga, B. J., *Gallium Nitride and Silicon Carbide Power Devices*, Singapore: World Scientific, 2017, p. 32.

[11] Mochizuki, K., et al., "Numerical Determination of Schottky Barrier Height of Nickel/n-Type Gallium Nitride Diodes Formed on Freestanding Substrate," *Journal of Modern Mathematics Frontier*, Vol. 3, No. 2, 2014, pp. 29–33.

[12] Maeda, T., et al., "Temperature Dependence of Barrier Height in Ni/n-GaN Schottky Barrier Diode," *Applied Physics Express*, Vol. 10, 2017, pp. 051002-1–051002-4.

[13] Mochizuki, K., et al., "Influence of Lateral Spreading of Implanted Aluminum Ions and Implantation-Induced Defects on Forward Current–Voltage Characteristics of 4H-SiC Junction Barrier Schottky Diodes," *IEEE Transactions on Electron Devices*, Vol. 56, No. 5, 2009, pp. 992–997.

[14] Pearton,, S. J, C. A. Abernathy, and F. Ren, *Gallium Nitride Processing for Electronics, Sensors and Spintronics*, London: Springer-Verlag, 2006, p. 44.

[15] Baliga, B. J., *Gallium Nitride and Silicon Carbide Power Devices*, Singapore, World Scientific, 2017, pp. 147–172.

[16] Baliga, B. J., *Silicon Carbide Power Device*, Singapore: World Scientific, 2005, p. 85.

[17] Kimoto, T., and J. A. Cooper, *Fundamentals of Silicon Carbide Technology*, Singapore: John Wiley & Sons, 2014, p. 249 and pp. 282–286.

[18] Crowell, C. R., and S. M. Sze, "Current Transport in Metal-Semiconductor Barriers," *Solid-State Electronics*, Vol. 9, No. 11–12, 1966, pp. 1035–1048.

[19] Itoh, A., T. Kimoto, and H. Matsunami, "High Performance of High Voltage 4H-SiC Schottky Barrier Diodes," *IEEE Electron Device Letters*, Vol. 16, No. 6, 1995, pp. 280–282.

[20] Suda, J., et al., "Nearly Ideal Current-Voltage Characteristics of Schottky Barrier Diodes Formed on Hydride-Vapor-Phase-Epitaxy-Grown GaN Freestanding Substrates," *Applied Physics Express*, Vol. 3, 2010, pp. 101003-1–101003-3.

[21] Tanaka, N., et al., "50 A Vertical GaN Schottky Barrier Diode on a Freestanding GaN Substrate with Blocking Voltage of 790V," *Applied Physics Express*, vol. 8, 2015, pp. 071001-1–071001-3.

[22] Fujiwara, H., et al., "Relationship Between Threading Dislocation and Leakage Current in 4H-SiC Diodes," *Applied Physics Letters*, Vol. 100, 2012, pp. 242102-1–242102-4.

[23] Johnson, J. W., et al., "Schottky Rectifiers Fabricated on Freestanding GaN Substrates," *Solid-State Electronics*, Vol. 45, No. 3, 2001, pp. 405–410.

[24] Zhang, A. P., et al, "Vertical and Lateral GaN Rectifiers on Freestanding GaN Substrates," *Applied Physics Letters*, Vol. 79, No. 10, 2001, pp. 1555–1557.

[25] Zhou, Y., et al., "High Breakdown Voltage Schottky Rectifier Fabricated on Bulk n-GaN Substrate," *Solid-State Electronics*, Vol. 50, No. 11–12, 2006, pp. 1744–1747.

[26] Zhou, Y., et al., "Temperature-Dependent Electrical Characteristics of Bulk n-GaN Schottky Rectifier," *Journal of Applied Physics*, Vol. 101, 2007, pp. 024506-1–024506-4.

[27] Ozbeck, A. M., and B. J. Baliga, "Planar Nearly Ideal Edge Termination Technique for GaN Devices," *IEEE Electron Device Letters*, Vol. 32, No. 3, 2011, pp. 300–302.

[28] Ozbeck, A. M., and B. J. Baliga, "Finite-Zone Argon Implant Edge-Termination for High-Voltage GaN Schottky Rectifiers," *IEEE Electron Device Letters*, Vol.

32, No. 10, 2011, pp. 1361–1363.

[29] Saitoh, Y., et al., "Extremely Low On-Resistance and High Breakdown Voltage Observed in Vertical GaN Schottky Barrier Diodes with High-Mobility Drift Layers on Low-Dislocation-Density GaN Substrates," *Applied Physics Express*, Vol. 3, 2010, pp. 081001-1–081001-3.

[30] Zhang, Q., et al., "12-kV p-Channel IGBTs with Low On-Resistance in 4H-SiC," *IEEE Electron Device Letters*, Vol. 29, No. 9, 2008, pp. 1027–1029.

[31] Bhatnagar, M, P. K. McLarty, and B. J. Baliga, "Silicon Carbide High Voltage (400 V) Schottky Barrier Diodes," *IEEE Electron Device Letters*, Vol. 13, No. 10, 1992, 501–503.

[32] Alok, D., B. J. Baliga, and P. K. McLarty, "A Simple Edge Termination for Silicon Carbide with Nearly Ideal Breakdown Voltage," *IEEE Electron Device Letters*, Vol. 15, No. 10, 1994, pp. 394–395.

[33] Miura, N., et al., "4H-SiC Power Metal–Oxide–Semiconductor Field Effect Transistors and Schottky Barrier Diodes of 1.7 kV Rating," *Japanese Journal of Applied Physics*, Vol. 48, 2009, pp. 04C085-1–04C085-4.

[34] Hontz, M. R., et al., "Modeling and Characterization of Vertical GaN Schottky Diodes with AlGaN Cap Layers," *IEEE Transactions on Electron Devices*, Vol. 64, No. 5, 2017, pp. 2172–2178.

[35] Chilukuri, R. K., and B. J. Baliga, "High Voltage Ni/4H-SiC Schottky Rectifiers," *International Symposium on Power Semiconductor Devices and ICs*, Toronto, May 26–28, 1999, pp. 161–164.

[36] McGlothlin, H. M., et al., "4 kV Silicon Carbide Schottky Diodes for High Frequency Switching Applications," *IEEE Device Research Conference*, Santa Barbara, June 28–30, 1999, pp. 42–43.

[37] Singh, R., et al., "SiC Power Schottky and PiN Diodes," *IEEE Transactions on Electron Devices*, Vol. 49, No. 4, 2002, pp. 665–672.

[38] Nakamura, T., et al., "A 4.15 kV 9.07 mΩ-cm² 4H-SiC Schottky Barrier Diode Using Mo Contact Annealed at High Temperature," *IEEE Electron Device Letters*, Vol. 26, No. 2, 2005, pp. 99–101.

[39] Mochizuki, K., et al., "Analysis of Leakage Current at Pd/AlGaN Schottky Barriers Formed on GaN Freestanding Substrates," *Applied Physics Express*, Vol. 4, No. 2, 2011, pp. 024104-1–024104-3.

[40] Ploog, K. H., and O. Brandt, "Doping of Group III Nitrides," *Journal of Vacuum Science and Technology A*, Vol. 16, No. 3, 1997, pp. 1609–1614.

[41] Ptak, A. J., et al., "Controlled Oxygen Doping of GaN Using Plasma Assisted Molecular-Beam Epitaxy," *Applied Physics Letters*, Vol. 79, No. 17, 2001, pp. 2740–2742.

[42] Hasegawa, H., and S. Oyama, "Mechanism of Anomalous Current Transport in n -Type GaN Schottky Contacts," *Journal of Vacuum Science and Technology B*, Vol. 20, No. 4, 2002, pp. 1647–1655.

[43] Nakamura, T., et al., "High Performance SiC Trench Devices with Ultra-low Ron," *International Electron Devices Meeting*, Washington, D.C., Dec. 5–7, 2011, pp. 599–601.

[44] Kimmering, L. C., "Recombination Enhanced Defect Reactions," *Solid-State Electronics*, Vol. 21, 1978, pp. 1391–1401.

[45] Hong, M. H., et al., "Stacking Fault Energy of 6H-SiC and 4H-SiC Single Crystals," *Phylosophical Magazine A*, Vol. 80, No. 4, 2000, pp. 919–935.

[46] Stampfl, C., and C. G. Van de Walle, "Energetics and Electronic Structure of Stacking Faults in AlN, GaN, and InN," *Physical Review B*, Vol. 57, No. 24, 1998, pp. R15052–R15055.

[47] Kajitani, R., et al., "A High Current Operation in a 1.6 kV GaN-Based Trenched Junction Barrier Schottky Diode," *Solid State Devices and Materials*, Sapporo, Sept. 27–30, 2015, pp. 1056–1057.

[48] Baliga, B. J., "The Pinch Rectifier: A Low Forward Drop High Speed Power Diode," *IEEE Electron Device Letters*, Vol. 5, No. 6, 1984, pp. 843–843.

[49] Held, R., N. Kaminski, and E. Niemann., "SiC Merged p-n/Schottky Rectifiers for High Voltage Applications," *Materials Science Forum*, Vol. 264–268, 1998, pp. 1057–1060.

[50] Dahlquist, F., et al., "Junction Barrier Schottky Diodes in 4H-SiC and 6H-SiC," *Materials Science Forum*, Vol. 264–268, 1998, pp. 1061–1064.

[51] Dahlquist, F., et al., "A 2.8 kV JBS Diode with Low Leakage," *Materials Science Forum*, Vol. 338–342, 2000, pp. 1179–1182.

[52] Wu, J., et al., "4,308 V, 20.09 mΩcm^2 4H-SiC MPS Diodes Based on a 30 Micron Drift Layer," *Materials Science Forum*, Vol. 457–460, 2004, pp. 1109–1112.

[53] Hull, B. A., et al., "Performance and Stability of Large Area 4H-SiC 10-kV Junction Barrier Schottky Rectifiers," *IEEE Transactions on Electron Devices*, Vol. 55, No. 8, 2008, pp. 1864–1870.

[54] Baliga, B. J., "Analysis of Junction-Barrier-Controlled Schottky (JBS) Rectifier Characteristics," *Solid-State Electronics*, Vol. 28, No. 11, 1985, pp. 1089–1093.

[55] Zhu, L., and T. P. Chow, "Analytical Modeling of High-Voltage 4H-SiC Junction Barrier Schottky (JBS) Rectifiers," *IEEE Transactions on Electron Devices*, Vol. 55, No. 8, 2008, pp. 1857–1863.

[56] Negoro, Y., T. Kimoto, and H. Matsunami, "Carrier Concentration Near Tail Region in Aluminum- or Boron-Implanted 4H-SiC (0001)," *Journal of Applied Physics*, Vol. 98, No. 4, 2005, pp. 043709-1–043709-7.

[57] Mochizuki, K., et al., "Influence of Lateral Spreading of Implanted Aluminum Ions and Implantation-Induced Defects on Forward Current–Voltage Characteristics of 4H-SiC Junction Barrier Schottky Diodes," *IEEE Transactions on Electron Devices*, Vol. 56, No. 5, 2009, pp. 992–997.

[58] Mochizuki, K., et al., "A Commercial-Simulator-Based Numerical-Analysis Methodology for 4H-SiC Power Device Formed on Misoriented (0001) Substrates," *IEEE Journal of the Electron Devices Society*, Vol. 3, 2015, pp. 316–322.

[59] Mochizuki, K., et al., "Uniform Luminescence at Breakdown in 4H-SiC 4Ω-off (0001) p–n Diodes Terminated with an Asymmetrically Spaced Floating-Field Ring," *IEEE Journal of the Electron Devices Society*, Vol. 3, 2015, pp. 349–354.

第 9 章

金属绝缘体半导体电容器和单极功率开关器件

9.1 引言

如 7.5.2 节所述，除了可以同时生产 CO 之外，还可以用与硅相同的方式对 SiC 进行热氧化。另一方面，通常可以沉积一层 SiN_x 或 Al_2O_3 层用作 GaN 和 AlGaN 的绝缘层（钝化层）（请参见 7.5.1 节）。因此，在本章中，金属绝缘体半导体（Metal Insulator Semiconductor，MIS）用于 GaN 和 SiC，而金属氧化物半导体（Metal Oxide Semiconductor，MOS）仅用于硅。

单极功率开关器件包括金属绝缘体半导体场效应晶体管（Metal Insulator Semiconductor Field-Effect Transistor，MISFET）和静电感应晶体管（Static Induction Transistor，SIT）[1]，它涵盖了 p-n 结栅极 SIT［也称为结型场效应晶体管（Junction Field-Effect Transistor，JFET）］，肖特基结栅极 SIT［也称为金属半导体场效应晶体管（Metal-Semiconductor Field-Effect Transistor，MESFET）］和一种 p^+ 栅极 SIT（也称为 p^+ 栅极 FET）。对于 JFET，0.6 kV 级 4H-SiC 产品可商购获得；然而，它们大多数具有常通特性（见图 9.1a），这阻碍了某些电路应用[2]。尽管传统的 MESFET 也具有常通特性，但据报道 V 形槽 AlGaN/GaN MESFET 具有常关特性（见图 9.1b）[3]。因此，本章将重点介绍 MISFET、MESFET 和 p^+ 栅极 FET。将简要回顾 MIS 电容器的物理原理，然后介绍 GaN 异质结场效应晶体管（Heterostructure Field-Effect Transistor，HFET）、4H-SiC JFET、GaN 和 4H-SiC MISFET，包括具有沟槽栅和超结结构的 MISFET［超结（Super

Junction，SJ）]（请参见 6.6 节和 7.2 节）。

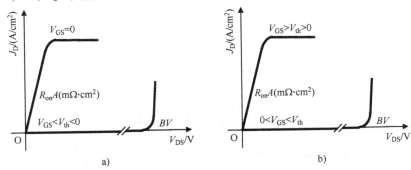

图 9.1　a）常开和 b）常关功率开关器件的漏极电流密度（J_D）漏极-源极电压（V_{DS}）的示意图。在 9.2.1 节中定义 V_{th} 为阈值电压

9.2　MIS 电容器

9.2.1　理想的 MIS 电容器

MIS 电容器是由金属、绝缘体和半导体层形成的电容器。尽管 MIS 电容器已直接应用于例如光学成像和信号处理的电荷耦合器件，但它不仅是 MISFET 的基本组成部分，还是研究半导体表面的最有用工具。

8.1 节考虑了金属/n 型半导体触点；相比之下，本节将研究金属/绝缘体（厚度：d）/p 型半导体结构。假设该结构的绝缘体中没有电荷，并且在绝缘体/p 型半导体界面处没有界面态或固定电荷（绝缘体和固定电荷对 MIS 结构的影响在 9.2.2 节中描述）。如 8.1 节所述，电子亲和能（qX_i 对绝缘体和 qX 对 p 型半导体的关系见图 9.2a）是导带底到真空中电子能级与导带能级最小值 E_c 的差。由于作为本征半导体的费米能级的本征费米能级 E_i 位于带隙 E_g 的中间，因此 p 型半导体的功函数 $q\Phi_s$ 表示为

$$q\Phi_s = qX + E_g/2 + q\Psi_F \tag{9.1}$$

式中，Ψ_F 为费米电势，定义为 E_i/q 与 E_F/q 之差，可以根据电离受主浓度 N_A^- 和本征载流子浓度 n_i 计算得出（参见 2.4 节）[4]

$$\Psi_F = (kT/q)\ln(N_A^-/n_i) \tag{9.2}$$

当金属功函数 $q\Phi_m$ 低于 $q\Phi_s$ 时，能带向下弯曲（见图 9.2b）。由于金属和

p型半导体的功函数之差 $q\Phi_{ms}$ 为

$$q\Phi_{ms}=q\Phi_m-qX-(E_g/2)-q\Psi_F \tag{9.3}$$

当对金属施加平带电压 $V_{FB}=\Phi_{ms}$ （<0）且 p 型半导体接地时，就实现了平带条件（见图9.2c）。

如果 p 型半导体保持接地，并且施加到金属上的电压 V 保持为负值，但幅度增加到 $|\Phi_{ms}|$ 以上，则表面电势 $|\Psi_S|$ 被定义为从衬底底部到表面的总带弯曲，变为负值（见图9.2d）。由于没有电流在 MIS 结构中流动，因此 E_F 保持平坦，从而导致在 p 型半导体表面附近聚集了空穴（所谓的堆积情况）。

另一方面，当 V 为正时，能带向下弯曲。当 V 太小以至于满足 $0<\Psi_S<\Psi_F$ 的条件时（见图9.2e），空穴从 p 型半导体的表面耗尽（所谓的耗尽情况）。与式（3.30）相似，耗尽层宽度 W_{Dp} 为

$$W_{Dp}=[2\varepsilon_r\varepsilon_o\Psi_S/qN_A]^{0.5} \tag{9.4}$$

耗尽区的总电荷 Q_D 可通过以下公式获得

$$Q_D=-qN_AW_{Dp}=-[2q\varepsilon_r\varepsilon_oN_A\Psi_S]^{0.5} \tag{9.5}$$

它被金属表面的正电荷相平衡。

当 V 足够大，以使 p 型半导体表面的 E_i 越过 E_F 时，Ψ_S 变为 $2\Psi_F$（见图9.2f）；即 p 型半导体表面的电子浓度等于本体的空穴浓度。发生该状态的电压称为阈值电压 V_{th}。当 $V>V_{th}$ 时被称为反型。一旦将 p 型半导体偏压于反型，p 型半导体表面的电子浓度就变成 Ψ_S 的近似指数函数，因此随 V 的增加而变化很小。所以，W_{Dp} 和整个耗尽区的总电势压降 $2\Psi_F$ 相对恒定。

图9.2 当金属功函数低于半导体功函数时，金属、绝缘体和 p 型半导体的能带图：a）分离，b）零偏压连接，c）平坦连接带偏置和偏置条件，d）累积，e）耗尽，f）反型

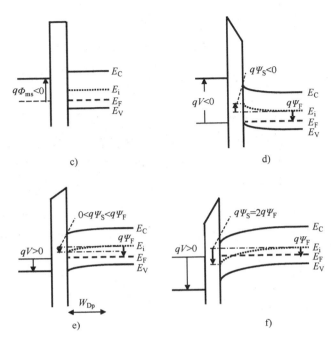

图 9.2 当金属功函数低于半导体功函数时,金属、绝缘体和 p 型半导体的能带图:
a) 分离,b) 零偏压连接,c) 平坦连接带偏置和偏置条件,d) 累积,e) 耗尽,f) 反型(续)

9.2.2 绝缘介质和固定电荷对 MIS 电容器的影响

首先对绝缘介质上电荷的影响进行了一维分析[5]。如图 9.3a 所示,考虑电荷密度 Q_1 位于平面 $x=x_1$ ($0 \leq x_1 \leq d$) 的情况。x_1 处的电荷会感应出相等且相反的电荷,这些电荷在 p 型半导体和金属之间分配。可以通过使用高斯定律来获得 V_{FB} (ΔV_{FB}) 的偏移。金属 ($x=0$) 和 Q_1 ($x=x_1$) 之间的恒定电场 E_1 为 (见图 9.3b):

$$E_1 = -Q_1/\varepsilon_1\varepsilon_0 \quad (9.6)$$

式中,ε_1 为绝缘介质的相对介电常数。ΔV_{FB} 由此获得

$$\Delta V_{FB} = x_1 E_1 = -x_1 Q_1/\varepsilon_1\varepsilon_0 \quad (9.7)$$

由于单位面积的绝缘介质电容 (C_1) 为

$$C_1 = \varepsilon_1\varepsilon_0/d \quad (9.8)$$

式 (9.7) 可以表示为

$$\Delta V_{FB} = -x_1 Q_1 / C_1 d \quad (9.9)$$

当 $x_1 = d$ 时达到最大值。

利用叠加和积分的方法计算分布在整个绝缘介质中的电荷所产生的增量,

通过电荷的任意分布 $\rho(x)$ 来概括 ΔV_{FB} 的上述方程式：

$$\Delta V_{FB} = -(1/C_I) \int_0^{x_I} (x/x_I) \rho(x) dx \qquad (9.10)$$

相反，绝缘介质/p 型半导体界面上的固定电荷（Q_f）通过 $-Q_f/C_I$ 贡献 V_{FB}。因此，包含 Φ_{ms}、绝缘介质电荷和界面电荷的影响因素的 ΔV_{FB} 可以写为

$$\Delta V_{FB} = \Phi_{ms} - (1/C_I) \int_0^{x_I} (x/x_I) \rho(x) dx - Q_f/C_I \qquad (9.11)$$

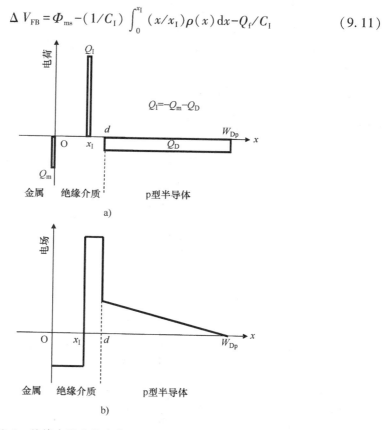

图 9.3 绝缘介质电荷密度 Q_I 对 MIS 结构的影响的示意图
a) 零偏压下的电荷配置（$Q_I = -Q_m - Q_D$），b) 零偏压下的电场分布

9.3 AlGaN/GaN 异质结构

图 9.4a 显示了在 GaN（0001）上形成的典型 AlGaN/GaN 异质结构的导带边缘 E_C 的分布。已知自发极化和压电极化都会在这种材料系统中感应电荷[6]。AlGaN 的较宽带隙会产生不连续性导带 ΔE_C，从而导致形成二维电子气

（2DEG）。如 8.8 节所述，与氧有关的浅施主（例如，对于 $Al_{0.26}Ga_{0.74}N$，能量深度 E_{SD} 为 0.03 eV[7,8]）存在于 AlGaN 表面，因此费米能级被固定在 E_{SD}。在 AlGaN 表面上，极化电荷 $-\sigma_{AlGaN}$ 使 qN_{SD}^+ 平衡，其中 N_{SD}^+ 是离子化的表面施主浓度（见图 9.4b）。由于在 AlGaN/GaN 界面上，2DEG 电荷 $-\sigma_{2DEG}$ 和极化电荷 $-\sigma_{GaN}$ 平衡极化电荷 $+\sigma_{AlGaN}$，因此 σ_{2DEG} 为

$$\sigma_{2DEG} = \sigma_{AlGaN} - \sigma_{GaN} \tag{9.12}$$

测量了在 GaN（0001）和 Si（111）衬底上形成的 2DEG 的薄层电子浓度 n_s 和电子迁移率 μ_n[9]。由于改善了 GaN 晶体的品质，$n_s^{GaN(0001)}$ 与 $n_s^{Si(111)}$（$5.6-5.7\times 10^{12}cm^{-2}$）相似，而 $\mu_n^{GaN(0001)}$（$1900cm^2/Vs$）比 $\mu_n^{Si(111)}$（$1450cm^2/Vs$）大 30%。该 2DEG 薄层电阻 $\rho_{2DEG}^{GaN(0001)}$ 为 $580\Omega/sq$，用于估计 9.5.1 节中的特征导通电阻 $R_{on}A$。

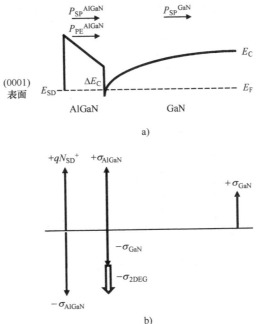

图 9.4 在 GaN（0001）晶面上形成的典型 AlGaN/GaN 异质结构
a）导带边缘，b）界面电荷的分布（P_{SP}：自发极化；P_{PE}：压电极化）

9.4 GaN、AlN、4H-SiC 和代表性绝缘介质的能带阵容

基于文献 [10-14]，图 9.5 总结了 GaN、AlN、4H-SiC 和代表性绝缘介质

的能带阵容。关于用于 GaN MIS 电容器的绝缘体，已经采用了 SiN_x[15]、Al_2O_3[16,17]和HfO_2[18]。据报道，Al_2O_3/（Al）GaN 界面处存在高密度的中间能隙态[19]，可能是由于形成了 Ga-O 键[20]。为了减少此类 Ga-O 键形成的影响，采用原位NH_3等离子体处理[21]，已经研究了原位 MOCVD 生长的薄 AlN 层[22,23]和原子层沉积的 AlN 层[24]。

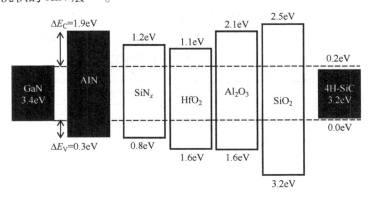

图 9.5　GaN、AlN、4H-SiC 和代表性绝缘介质的能带阵容[10-14]

9.5　GaN HFET

氮化镓异质结场效应晶体管（GaN HFET）是在 2DEG 沟道上外延生长宽禁带材料层（即 AlGaN）的 GaN FET。由于 2DEG 的高电子迁移率，它也被称为 GaN HEMT（请参见 9.3 节）。宽禁带材料可以用作多个绝缘介质层的一部分（即 GaN MIS HFET，请参见 9.5.1 节），也可以用作单个绝缘介质层（即 GaN MESFET 和 GaN P^+栅极 HFET，请参见 9.5.2 节和 9.5.3 节）。

9.5.1　GaN MIS HFET

在 GaN MISFET 中，在 AlGaN 层上形成绝缘介质层，以抑制栅极泄漏电流。首个在独立 GaN 衬底上的垂直型 GaN MIS HFET 的横截面示意图如图 9.6 所示[25]。电子电流流经 AlGaN/GaN 异质结和槽，并以大约 45°角扩散到漂移层中，并最终变得均匀。在截止状态期间，P^+-GaN 区起到电流阻挡区的作用。根据式（3.25），耗尽的P^+-GaN 区域中受主的薄层浓度等于耗尽的 n^--GaN 漂移层中的施主的薄层浓度。即使耗尽的 n^--GaN 宽度等于 n^--GaN 漂移层的厚度（t_{drift}），耗尽的 P^+-GaN 宽度也必须小于 P^+-GaN 区的厚度（t_{p+}），以防止发生称为穿通的现象。在文献[25]中满足此条件的原因是，只要掺杂的镁原子和硅原子分别成为受主和施主，p^+-GaN 区域中的薄层受主浓度（$3 \times 10^{14} cm^{-2}$）比 n^--GaN 薄层施主浓度（$3 \times 10^{12} cm^{-2}$）大得多。然后可以将击穿电压 BV（见图 9.1）确定为 P^+/n GaN 结处的最大电场（3.5 节中的 E_{max}）等于临界电场（2.6 节中的 $E_{critical}$）。然而，在文献[25]中没有报道击穿电压 BV。

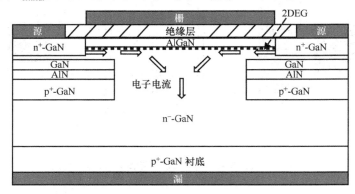

图 9.6 首个在独立 GaN 衬底[25]上的垂直型 GaN MIS HFET 的横截面示意图

图 9.7 所示为形成槽的工艺流程。在 MOCVD 工艺制备的 n^+-GaN（0001）衬底上（参见 6.2 节），生长了 $3\mu m$ 厚的 n^--GaN 层，$0.1\mu m$ 厚的 p^+-GaN 层，10nm 厚的 AlN 层和 50nm 厚的 GaN 层（见图 9.7a）。GaN/AlN/p^+-GaN 层通过 ICP 干法刻蚀，采用 Cl_2 和 SiO_2 掩模进行干法刻蚀（请参见 7.2.1 节）（见图 9.7b）。去除掩模后，通过以下步骤生长厚度为 $0.3\mu m$ 的 n^--GaN 层（Si：$1 \times 10^{16} cm^{-3}$）和通过 MOCVD 生长厚度为 15nm 的 $Al_{0.25}Ga_{0.75}N$ 层（见图 9.7c）。这里 AlN 层用于抑制从表面到槽区域的质量输运[26]。

图 9.7 图 9.6[25]中所示槽的制备的工艺流程

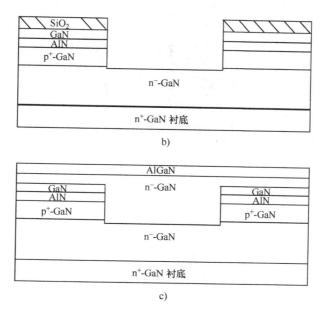

图 9.7 图 9.6[25]中所示槽的制备的工艺流程（续）

硅离子注入用于形成 n^+-GaN 源区（请参见 7.3.1 节）。作为栅绝缘层和电极，通过低压化学气相沉积来沉积厚度为 50nm 的高温 SiO_2 层和厚度为 250nm 的磷掺杂的多晶硅膜。退火以激活多晶硅（在 850℃ 下保持 20min），同时也激活了氢化镁[27]。应该注意的是，在其上生长 n^--GaN 的 p^+-GaN 中的镁不会通过激活退火而完全脱氢[28]；所以，最近使用氨分子束外延代替 MOCVD 来制造 GaN MIS HFET，而没有进行激活退火[29]。

在文献 [25] 中，通过电子束蒸发形成由钛（20nm）/铝（1μm）制成的源电极和漏电极。GaN MIS HFET 具有 V_{th} 为 -16V 的常开特性。栅极-源极电压 V_{GS} 为 0V 时，$R_{on}A$ 为 2.6mΩ·cm²，可以通过减小 JFET 区域 W_{JFET} 的宽度进一步减小（见图 9.8）。

理想情况下，$R_{on}A$ 由图 9.8 所示电流路径中组件的特征电阻确定：

$$R_{on}A = R_S A + R_{ch} A + R_{accum} A + R_{JFET} A + R_{drift} A + R_{sub} A \quad (9.13)$$

式中，$R_S A$ 为源电阻；$R_{ch} A$ 为沟道电阻；$R_{accum} A$ 为累积区电阻；$R_{JFET} A$ 为 JFET 区电阻；$R_{drift} A$ 为漂移区电阻；$R_{sub} A$ 为衬底基片电阻（请参阅 1.3.1 节）。

在式（9.13）中，$R_S A$ 可忽略不计，因为源区通常是重掺杂的。$R_{ch} A$ 给出公式为

$$R_{ch} A = \rho_{2DEG}^{GaN(0001)} L_{ch}(P/2) \quad (9.14)$$

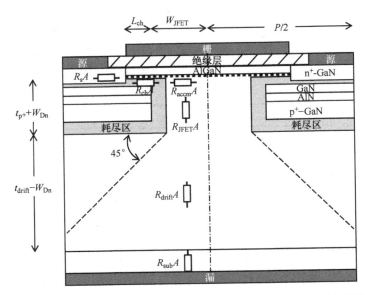

图 9.8　图 9.6 所示的垂直型 GaN MIS HFET 的电流路径中的特征电阻

式中，L_{ch} 为沟道长度；P 为单元节距（见图 9.8）。$R_{ch}A$ 可以近似为

$$R_{accum}A = 0.6\rho_{2DEG}^{GaN(0001)} W_{JEFT}(P/2) \tag{9.15}$$

其中系数 0.6 通常用于垂直型硅功率开关器件，以解决从沟道到 JFET 区域的二维电流扩散[30]。$R_{JFET}A$、$R_{drift}A$、$R_{sub}A$ 表示为

$$R_{JFET}A = \rho_{drift}(t_{p^+}+W_{Dn})(P/2)/(W_{JEFT}-W_{Dn}) \tag{9.16}$$

$$R_{drift}A = \rho_{drift}\{[(P/2)\ln(P/2)/(W_{JEFT}-W_{Dn})]+t_{drift}-W_{Dn}-[(P/2)-W_{JEFT}]\} \tag{9.17}$$

$$R_{sub}A = \rho_{sub} t_{sub} \tag{9.18}$$

式中，ρ_{drift} 为漂移区的电阻率；ρ_{sub} 为衬底的电阻率；t_{sub} 为衬底的厚度，耗尽的 n^--GaN 宽度 W_{Dn} 已经在式（3.30）中给出，即

$$W_{Dn} \approx (2\varepsilon_r\varepsilon_o\Psi_{bi}/qN_D)^{0.5} \tag{9.19}$$

注意，在式（9.13）中，由于其值较小（即 $8×10^{-5}\Omega\ cm^2$）而忽略了特征接触电阻[25]。内置电势 Ψ_{bi} 给出为[31]

$$\Psi_{bi} = (KT/q)\ln(N_A N_D/n_i^2) \tag{9.20}$$

其中 n_i^{GaN} 已经在式（2.7a）中给出，即

$$n_i^{GaN} = 2.0×10^{15} T^{1.5} \exp(-2.0×10^4/T)(cm^{-3}) \tag{9.21}$$

在 $P/2 = W_{JEFT}+4\mu m$ 的情况下，垂直型 GaN MIS HFET 在 300K 时的 $R_{on}A$

是 W_{JFET}（见图 9.9）。在文献 [25] 中，$L_{ch} = 2\mu m$，$W_{JFET} = 1.5\mu m$。用于计算曲线的其他参数在图 9.9 中列出如下：$\rho_{2DEG}^{GaN(0001)} = 580\Omega/sq$（见 9.3 节）；$\mu_n = 850\ cm^2/Vs$（见图 2.16）；$\varepsilon_r = 10.4^{[32]}$；$\rho_{sub} = 0.018\Omega cm$；$t_{sub} = 600\mu m^{[33]}$。在 $W_{JFET} = 3.5\mu m$ 时计算得出的 $R_{on}A$ 最小为 $1.7m\Omega \cdot cm^2$。当 $W_{JFET} < 3.5\mu m$ 时，$R_{accum}A$ 随着 W_{JFET} 的增加而增加，而当 $W_{JFET} > 3.5\mu m$ 时，$R_{drift}A$ 和 $R_{JFET}A$ 的贡献变大。由于 $R_{sub}A$（即 $1.1m\Omega \cdot cm^2$）的贡献占主导地位，因此在 $3\mu m$ 厚的 $1 \times 10^{16} cm^{-3}$ 掺杂的 n^--GaN 漂移层的情况下，减薄 GaN 衬底是降低 $R_{on}A$ 最有效的措施。

注意，与上述外延生长的 p^+ 型 GaN 掩埋区相反[25]，还使用了注入镁离子的 p^+ 型 GaN 掩埋区[34]。在 $NH_3 + N_2$ 环境中，于 1553K 进行激活退火，得到的 BV 和 $R_{on}A$ 分别为 250V 和 $2.2m\Omega \cdot cm^2$。

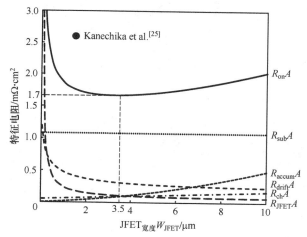

图 9.9 在 $P/2 = W_{JFET} + 4\mu m$ 的情况下，垂直型 GaN MIS HFET 的特征电阻（见图 9.6）与 JFET 宽度的函数关系，由式（9.13）~式（9.21）计算得出

9.5.2 GaN MESFET

在 GaN MESFET 中，在 AlGaN 层上没有形成绝缘体层。如图 9.10 所示，自发极化的方向平行于 GaN（11$\bar{2}$0）表面。由于仅压电极化会在 GaN（11$\bar{2}$0）上形成的 AlGaN/GaN 界面处感应电荷，所以 n_s 小于在 GaN（0001）上形成的 AlGaN/GaN 界面处的电荷。如果在（0001）和（11$\bar{2}$0）之间的面上形成 AlGaN/GaN 界面，则 n_s 取 $n_s^{(0001)}$ 和 $n_s^{(11\bar{2}0)}$ 之间的值。通过 MOCVD[3] 制备的 n^+-GaN

(0001) 衬底，在衬底上依次生长厚度为 0.2μm 的 n$^+$-GaN 层（Si：$3×10^{18}$cm^{-3}，厚度为 1μm 的 p$^+$-GaN 层（Mg：$5×10^{18}$cm^{-3}）和 5μm 厚的 n$^-$-GaN 层（Si：$7×10^{15}$cm^{-3}）。通过 ICP 干法刻蚀在 n$^+$-GaN/p$^+$-GaN/n$^-$-GaN 层（倾斜角为 16°）中形成 V 形沟槽，并重新生长形成 75nm 厚的 Al$_{0.25}$Ga$_{0.75}$N/GaN 层（见图 9.11）。通过将 Al$_{0.25}$Ga$_{0.75}$N 层从 35 nm 减薄到 10 nm，V_{th} 从 -3.2V 变为 +0.3V。对于常开型器件，报道的 BV 为 672V，$R_{on}A$ 为 7.6mΩ·cm^2。但是，对于常关型器件，均未对 BV 和 $R_{on}A$ 进行相关的报道。

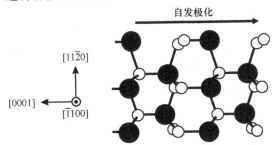

图 9.10　GaN（11$\bar{2}$0）的示意性截面图，显示了自发极化的方向
（实心圆：镓原子；空心圆：氮原子）

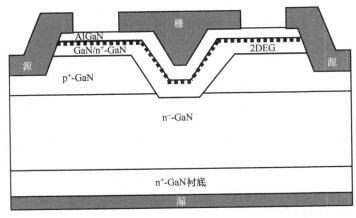

图 9.11　在独立式 GaN 衬底[3]上的垂直型 GaN MESFET 的横截面示意图

9.5.3　GaN p$^+$ 栅极 HFET

在 GaN p$^+$ 栅极 HFET 中，在 AlGaN 层上形成 p$^+$ 型 GaN 层。通过适当设计的 p$^+$ 型 GaN 层，异质结的电势仅在栅极下方升高，从而产生正 V_{th}（见图 9.12 中的实线）。

在n+型GaN衬底上生长15μm厚，掺杂浓度为$1×10^{16}$cm^{-3}的n型GaN漂移层（见图9.13a），并得到BV和$R_{on}A$分别为1.5 kV和2.2mΩ·cm^2；然而，V_{th}相对较低（即0.5V）[35]。

根据（0001）的表面倾斜角，计算出p+-GaN/AlGaN/GaN的n_s[36]（见图9.14）。当在n型GaN漂移层上方形成的V形槽上方重新生长p+-GaN/AlGaN/GaN层时，通过降低n_s可使V_{th}增加到2.5V。

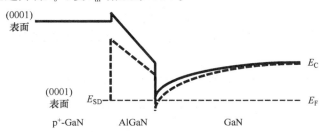

图9.12 在GaN（0001）上以0 V的栅极-源极偏置形成的p+-GaN/AlGaN/GaN（实线）和AlGaN/GaN（虚线）异质结构的导带边缘分布

根据这一发现，在p+-GaN（Mg：$3×10^{19}$cm^{-3}）和13μm厚的n−-GaN（Si：$1×10^{16}$cm^{-3}）漂移层的V形槽表面通过MOCVD方法再生长p+-GaN/AlGaN/GaN层（见图9.13b）。还插入了一个绝缘的GaN层，掺杂了$5×10^{18}$cm^{-3}的碳，以阻挡截止态泄漏电流。通过ICP干法刻蚀选择性腐蚀p+-GaN形成栅极和AlGaN/GaN/carbon掺杂的GaN形成源电极后，在前侧形成钛/铝源极和钯/金栅电极，并在背面形成钛/铝/钛/金漏电极。所得的BV和$R_{on}A$分别为1.7 kV和1.0mΩ·cm^2。展示了稳定的栅极特性并成功实现了400V/15A快速切换。

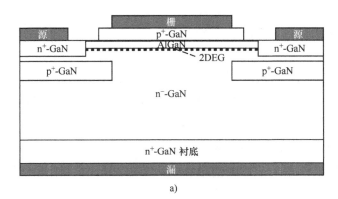

图9.13 独立GaN衬底上的a）平面[35]和b）V形槽[36] GaN p+栅极HFET的示意性剖视图

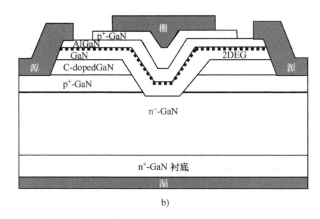

图 9.13 独立 GaN 衬底上的 a）平面[35] 和 b）V 形槽[36] GaN p+栅极 HFET 的示意性剖视图（续）

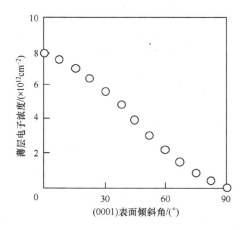

图 9.14 计算得出 p+-GaN/AlGaN/GaN 沟道的薄层电子浓度与（0001）表面倾斜角的关系[36]

9.6 4H-SiC JFET

JFET 是掩埋沟道形式的 FET，因此没有表面效应。但是，栅极与通道之间的距离较大，因此实现常关特性具有挑战性（请参见 9.1 节）。基于沟槽和注入垂直沟道结构（见图 9.15a）的常关型 4H-SiC JFET 的 $R_{on}A$ 为 3.6mΩ·cm^2，BV 为 1.7kV[37] 和 $R_{on}A$ 为 130mΩ·cm^2，BV 为 11 kV[38]。但是，狭窄的通道宽度（小于 1.7μm[37] 和 0.55μm[38]）对光刻提出了严格的要求。通过在栅极和漏极之间插入 p+屏蔽栅，可实现 $R_{on}A$ 为 2.4mΩ·cm^2 和 BV 为 1.4 kV，但 V_{th} 则低至+ 1.0V[39]。

图 9.15b 中展示了一种具有埋入式 p+栅作为控制栅极的 4H-SiC JFET，其 $R_{on}A$ 为 1.0mΩ·cm^2[40]。但是，在正常情况下，在 V_{GS}=-12V 时测得的 BV 为 700V。

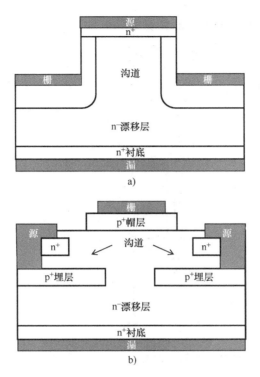

图 9.15 具有 a）沟槽与注入沟道和 b）p^+ 埋栅的 4H-SiC JFET 的示意截面图

当高压常开 4H-SiC JFET 与低压常关硅 MOSFET 级联时，从外部角度看，JFET-MOSFET 对用作常关功率开关器件（见图 9.16）[41]。最近用 80V 常关硅 MOSFET 级联了一个 3.3kV 常开 4H-SiC JFET，其阻断电压高于 4.0kV，$R_{on}A$ 低至 14.7mΩ·cm^2 [42]。

9.7 MISFET

MISFET 是包括 MIS 电容器、漂移层以及源极和漏极区域的 FET。MIS 电容器中的栅极结构分为平面型和沟槽型，而漂移层结构分为常规的 n 型和 SJ 型。图 9.17a ~ c 所示分别为平面、沟槽和 SJ MISFET 的示意截面图。平面 Si MOSFET 是第一个商用成功的单极功率开关器件。为了降低制造成本，这些器件中的沟道是通过所谓的双扩散工艺形成的。即，p 型和 n 型掺杂剂的横向扩散率之间的差异限定了沟道。另外，对于 GaN 和 4H-SiC MISFET，由于缺乏掺杂剂扩散（例如 GaN 中的硅和镁，4H-SiC 中的铝）和复杂的掺杂剂扩散（例如 4H-SiC 中的硼），致使 SiC 无法使用双扩散工艺（请参阅 7.4 节）。因此，GaN 和 4H-SiC 平面 MISFET 通过外延生长和/或离子注入工艺进行制备。

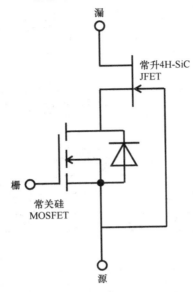

图 9.16 高压常开 4H-SiC JFET 与低压常关硅 MOSFET 级联

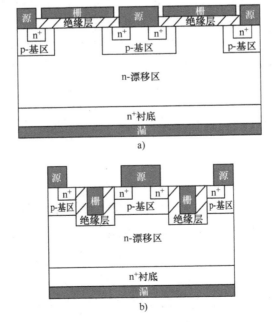

图 9.17 a) 平面，b) 沟槽和 c) SJ MISFET 的示意截面图

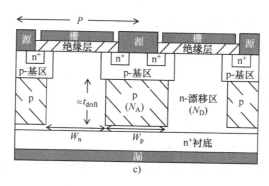

图 9.17 a）平面，b）沟槽和 c）SJ MISFET 的示意截面图（续）

沟槽 MISFET（由于栅极绝缘体的 U 形横截面，也称为 UMISFET）（见图 9.17b）可以增加沟道密度，因为 MIS 沟道垂直于表面。此外，相比平面 MISFET，沟槽 MISFET 可以消除 $R_{JFET}A$，获得极低的 $R_{on}A$[43]。

SJ 器件[44]的概念首先应用于变容二极管[45]，其中交替的 p 型和 n 型柱位于漂移层中，然后应用于硅 SJ MOSFET 的商业化生产[46]。尽管图 9.17c 显示了平面 SJ MISFET，但也可以制造沟槽 SJ MISFET。

9.7.1 平面 MISFET

据作者所知，尚无平面 GaN MISFET 的报道。在 SiC 的应用中，1997 年使用双注入工艺制造了平面 6H-SiC MISFET[47]，其中将多种能量的硼离子注入以形成 1μm 深的 p 基区（见图 9.17a）。MIS 结构中使用的绝缘介质层是热生长的氧化物，并且反型层沟道迁移率 μ_{inv} 为 20 cm^2/Vs。报道的 BV 和 $R_{on}A$ 分别为 760V 和 130mΩ·cm^2。

在 2001 年，使用沉积的氧化物制造了平面 4H-SiC MISFET[48]。报告的 μ_{inv}，BV 和 $R_{on}A$ 分别为 14 cm^2/Vs，2 kV 和 55mΩ·cm^2。关于 BV 和 $R_{on}A$，到 2004 年已报道了改进的值，例如 2.4 kV 和 42mΩ·cm^2[49]与 10 kV 和 236mΩ·cm^2[50]。2015 年，人们展示了平面 4H-SiC MISFET 的 $R_{ch}A$ 的改进，其中在沟道区中形成沟槽以增加单位面积的沟道密度[51]。

另一方面，通过使用氮气改善 MIS 界面质量[53]（见 7.5.2 节），μ_{inv} 随着时间的推移一直在增加（例如高达 130 cm^2/Vs[52]）或在后氧化过程中掺杂磷[54,55]；锶[56]，钡[57]，锑[58]或镧[52]钝化界面电荷陷阱，并在干燥氧化物中扩散悬空键的硼钝化[59,60]。μ_{inv} 限制因素包括库仑散射、表面散射和声子散射[61]；

对于$SiO_2/4H\text{-}SiC$界面，Noguchi 等人通过实验观察到声子极限μ_{inv}，并得出结论，在高电压下，表面粗糙度不是μ_{inv}的主要限制因素[62]。

当V_{GS}超过V_{th}时，p 基和 JFET 区域的表面分别对应于反型和累积情况（见 9.2.1 节）。电子电流流经反转层沟道（在 p 基极区域的表面上形成）进入积累层（在 JFET 区域的表面上形成），并以大约45°角扩散到漂移层中，最终变得统一（见图 9.18a）。为了减少$R_{JFET}A$，已使用了重掺杂的 n 型 JFET 区域和电流扩散层（CSL）（见图 9.18b）[63]。使用这种结构时，理想情况是由电流路径中组件的特征电阻确定平面 4H-SiC MISFET 的$R_{on}A$（见图 9.18b），如下所示：

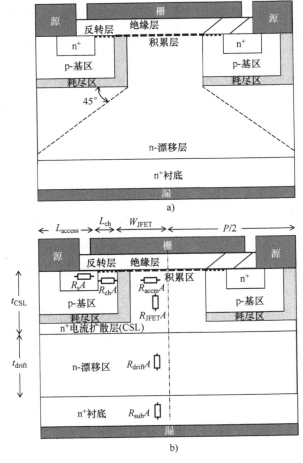

图 9.18　a）没有 CSL 的平面 4H-SiC MISFET 的电流路径，以及 b）有 CSL 的平面 4H-SiC MISFET 的电流路径中的电阻

$$R_{on}A = R_S A + R_{ch}A + R_{accum}A + R_{JFET}A + R_{drift}A + R_{sub}A \tag{9.22}$$

在式（9.22）中，$R_S A$ 可忽略不计，因为源区通常是重掺杂的。$R_{ch}A$ 给出为

$$R_{ch}A = L_{ch}(P/2)/[\mu_{inv}C_I(V_{GS}-V_{th})] \tag{9.23}$$

式中，C_I 为栅极绝缘体的特征电容，表示为电容率与栅极绝缘体厚度的比值；即

$$C_I = \varepsilon_I/t_I \tag{9.24}$$

$R_{accum}A$ 给出为

$$R_{accum}A = 0.6(W_{JFET}-W_{Dn})(P/2)/[\mu_{accum}C_I(V_{GS}-V_{th})] \tag{9.25}$$

式中，μ_{accum} 为累积层中的电子迁移率。这里在式（9.15）中已经出现的 0.6 因子通常用于垂直型硅功率开关器件，以解决二维电流从沟道扩散到 JFET 区域[30]的问题。使用式（2.7b）获得 W_{Dn}。

$$n_i^{4H-SiC} = 3.9 \times 10^{15} T^{1.5} \exp(-1.9 \times 10^{14}/T)(\text{cm}^{-3}) \tag{9.26}$$

式（9.19）和式（9.20）中的 $R_{JFET}A$ 和 $R_{drift}A$ 可以分别近似为

$$R_{JFET}A = \rho_{CSL}t_{CSL}(P/2)/(W_{JFET}-W_{Dn}) \tag{9.27}$$

$$R_{drift}A = \rho_{drift}t_{drift} \tag{9.28}$$

式中，ρ_{CSL} 和 t_{CSL} 分别为 CSL 的电阻率和厚度。$R_{sub}A$ 由式（9.18）给出。

在文献 [64] 中报道的双自对准短沟道 4H-SiC 平面 1kV 级 MISFET 的情况下，$L_{ch} = 0.5\mu m$；$P/2 = 4\mu m$；$t_I = 50nm$；$W_{JFET} = 0.5\mu m$；$N_{CSL} = 1.0 \times 10^{17} \text{cm}^{-3}$；$t_{CSL} = 4\mu m$；$N_{drift} = 8.55 \times 10^{15} \text{cm}^{-3}$；$t_{drift} = 8.4\mu m$；$V_{GS}-V_{th} = 20V$。假设 $\mu_{accum} = \mu_{inv}$，$\mu_{CSL} = 600 \text{cm}^2/\text{Vs}$，$\mu_{drift} = 850 \text{cm}^2/\text{Vs}$（见图 2.16），$\rho_{drift} = 0.02\Omega \cdot \text{cm}^{[65]}$ 和 $t_{drift} = 350\mu m^{[65]}$，300K 条件下的 $R_{on}A$ 计算为 μ_{inv} 的函数（见图 9.19）。由于使用 CSL，$R_{JFET}A$ 可以忽略不计。尽管在文献 [64] 中未描述 μ_{inv}，但图 9.19 的结果表明，在 4H-SiC 平面 1kV 级 MISFET 的情况下，测得的 $R_{on}A$ 为 $6.7m\Omega \cdot \text{cm}^{2[64]}$，应随着 μ_{inv} 的增加而减小。

对于更高电压的 4H-SiC 平面 MISFET，BV 为 1.7kV、3.3kV 和 4.5kV 的漂移层中典型厚度/施主浓度分别为 $15\mu m/2 \times 10^{15} \text{cm}^{-3}$、$34\mu m/1.5 \times 10^{15} \text{cm}^{-3}$ 和 $40\mu m/1 \times 10^{15} \text{cm}^{-3[66]}$。在三种掺杂水平下，假设 μ_{drift} 为 $850 \text{cm}^2/\text{Vs}$（见图 2.16），则将 300K 条件下的 $R_{on}A$ 计算为 μ_{inv} 的函数（见图 9.20）。由于在这种高压 MISFET 中 $R_{drift}A$ 占 $R_{on}A$ 的主导地位，因此就减小 $R_{on}A$ 而言，减小 μ_{inv} 并不那么重要。

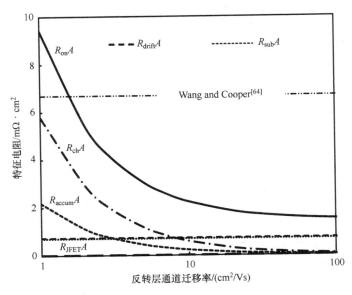

图 9.19 具有 CSL 的 4H-SiC 平面 1kV 级 MISFET 的特征电阻（见图 9.16b），由式（9.18）~式（9.20）和式（9.22）~式（9.28）计算得到

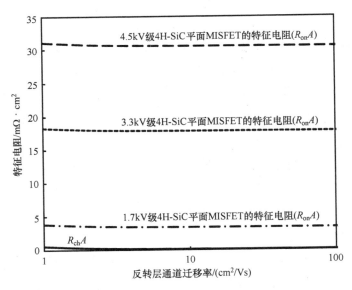

图 9.20 具有 CSL 的 4H-SiC 平面 1.7~4.5kV 级 MISFET 的特征电阻（见图 9.18b）与由式（9.18）~式（9.20）和式（9.22）~式（9.28）得到的关于式（9.28）中的反型层沟道迁移率的函数

9.7.2 沟槽 MISFET

沟槽 MISFET 的基本结构如图 9.17b 所示。自从 1985 年报道了第一个 Si 沟槽 MOSFET[43]以来，第一个 4H-SiC[67]和 GaN[68]沟槽 MISFET 的出现分别经历了 11 年和 23 年。

9.7.2.1 GaN 沟槽 MISFET

独立 GaN 衬底上 GaN 沟槽 MISFET 的 BV 和 $R_{on}A$ 的数据在图 9.21[68-76]中列出。关于栅极布局，使用单元节距为 12.6μm 的规则六边形沟槽栅极（见图 9.22）时，栅极宽度与单位单元面积之比（W_G/S_{cell}）为 0.267μm^{-1}[71]，是条纹布局的 W_G/S_{cell} 的两倍，单元节距为 15μm[70]。对于（1.5×1.5）mm^2 芯片（有效面积：1.8mm^2），六边形布局的单元节距为 18μm，从而产生了大电流（23.2A）和快速开关特性[72]。然而，常规沟槽 MISFET 的反向 p-GaN 沟道中的 μ_{inv} 很低。为了增加沟道中的电子迁移率，报道了采用合适的氧化物的 GaN 中间层为基础的垂直沟道 MOSFET（OG-FET）[73,74]和 GaN 垂直场效应晶体管[75,76]。前者是使用 GaN 中间层的再生长然后在沟槽结构上进行 MOCVD 电介质沉积来制造的，而后者仅使用具有亚微米鳍形沟道的 n 型 GaN 层。

9.7.2.2 4H-SiC 沟槽 MISFET

尽管 4H-SiC MISFET 的 BV 与第一个 4H-SiC 沟槽 MISFET 的 BV（即 0.26 kV[67]）在 1997 年[77]和 1998 年[78]分别增加到 1.1 kV 和 1.4 kV，但在 4H-SiC 整个栅极绝缘体上产生的非常强的电场限制了 4H-SiC 沟槽 MISFET 的 BV。

作者及参考文献	Otake et al.[68]	Kodama et al.[69]	Oka et al.[70]	Oka et al.[71]	Oka et al.[72]
年份	2008	2008	2014	2015	2016
击穿场强/kV	–	0.18	1.6	1.25	1.6
特征导通电阻/mΩ·cm^2	9.3	–	12.1	1.8	2.7

作者及参考文献	Gupta et al.[73]	Gupta et al.[74]	Sun et al.[75]	Zhang et al.[76]
年份	2017	2017	2017	2017
击穿场强/kV	0.99	1.4	0.8	1.2
特征导通电阻/mΩ·cm^2	2.6	2.2	0.36	0.2

图 9.21 独立 GaN 衬底上的 GaN 沟槽 MISFET 的 BV 和 $R_{on}A$ 的数据

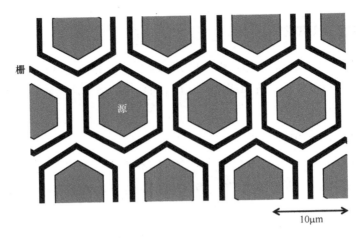

图 9.22 规则六边形沟槽栅布局[71,72]的平面示意图

即使在使用硅沟槽 MOSFET 的情况下,在硅沟槽底部拐角处的栅氧化层中也会产生强电场[79];但是,底层半导体中的电场E_s很弱,为 0.3MV/cm[80]。另一方面,E_s 会变高,即 3.75MV/cm[81] 用于 GaN,2.50MV/cm 用于 4H-SiC[82](请参见 2.6 节)。根据高斯定律,栅极绝缘层中的电场(E_1)与E_s的关系如下

$$E_1 = (\varepsilon_r/\varepsilon_1)E_s \qquad (9.29)$$

式中,ε_r 和 ε_1 分别为半导体和绝缘层的相对介电常数。

由于平行于 GaN 的 c 轴的 ε_r 为 10.4[83],而 6H-SiC 的 ε_r 为 10.0[84](参见 2.7 节),对于Si_3N_4和SiO_2,其ε_1分别为 7.5 和 3.9[85]。E_1变大,即在 GaN 上使用Si_3N_4时E_1为 5.2 MV/cm,在 4H-SiC 上使用SiO_2时E_1为 6.4MV/cm[67]。这些E_1值接近于材料本身的绝缘介电强度(对于Si_3N_4和SiO_2[85]均约等于 10MV/cm),导致在使用高V_{DS}时沟槽底部栅绝缘层被破坏[67]。

因此,屏蔽如此高的E_1是必不可少的;然而,据作者所知,尚未发现 GaN 屏蔽沟槽 MISFET 的相关报道。相反,在 2002 年,P^+屏蔽区域被短路到源极端,首次将源极并入 4H-SiC 沟槽 MISFET 的沟槽底部(见图 9.23a)[86]。对于这样的 4H-SiC 屏蔽沟槽 MISFET 的 BV,掺杂了 50μm 厚度的漂移层的 BV 增加到 3kV,掺杂厚度为 115μm 的漂移层的浓度 $8.5\times10^{14}cm^{-3}$[86]和 115μm 厚度的漂移层的 BV 增加到 5 kV 漂移层的浓度 $7.5\times10^{14}cm^{-3}$[87]。但是,据报道,这种类型的 4H-SiC 沟槽 MISFET 的$R_{on}A$ 值相对较大,一些文献上有相关报道,$R_{on}A$ 值为:18mΩ·cm²[67],180mΩ·cm²[77],311mΩ·cm²[78],120mΩ·cm²[86]和 228mΩ·cm²[87]。

2011 年,据报道带有源极和栅极沟槽的 4H-SiC MISFET[所谓的双沟槽(见

图9.23b)] 具有非常低的$R_{on}A$,在$BV = 630V$时$R_{on}A$为:$0.79 mΩ·cm^2$($N_{drift} = 1.8×10^{16} cm^{-3}$;$t_{drift} = 5μm$);在$BV = 1260V$时$R_{on}A$为:$1.41 mΩ·cm^2$($N_{drift} = 7.5×10^{15} cm^{-3}$;$t_{drift} = 8μm$)[88]。如8.8节所述,这种双沟槽结构对于减小沟槽SBD中的电场也是有效的。必须注意的是,在文献[88]中,t_{sub}变薄至100μm。在双沟槽MISFET的沟槽侧壁上测得的$μ_{inv}$相对较小(即11 cm^2/Vs)[88]。据Harada等报道[89],深P^+区域在沟槽下方创建了一个JFET,势垒减小了沟槽底部的电场,如果深P^+区域的形成比栅极沟槽的底部深1.2μm,则E_I可以减小至小于3 MV/cm。

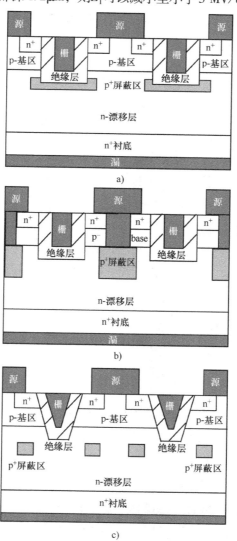

图9.23 P^+屏蔽的 a)单沟槽[79],b)双沟槽[88]和c)V沟槽MISFET[90]的示意截面图

9.7.2.3 4H-SiC V 沟道沟槽 MISFET

在沟道区域中使用 4H-SiC（$0\bar{3}3\bar{8}$）面的 V 形槽 MISFET 如图 9.23c 所示[90]。它的制造过程开始于生长一个厚度为 12μm 的 n 型漂移层，掺杂浓度为 $4.5×10^{15}cm^{-3}$。然后，通过铝离子注入形成 P^+ 屏蔽区域，生长 3μm 厚的 $7×10^{15}$ cm^{-3} 掺杂的 n 型漂移层。为了在沟槽侧壁上形成长为 0.6μm 的沟道，分别用铝和磷注入 p 基和 n^+ 源区，BV 和 $R_{on}A$ 分别为 1700V 和 3.6mΩ·cm^2，而对于不带 p^+ 屏蔽区的 V 形槽 MISFET，BV 和 $R_{on}A$ 分别为 575V 和 3.1mΩ·cm^2。据报道，沿着沟槽侧壁的 μ_{inv} 为 80 cm^2/Vs[91]，其中 P^+ 区域占据有效面积 30%的设计平衡了 BV 和 $R_{on}A$ 的综合性能[92]。

9.7.3 SJ MISFET

根据高斯定律，给定的 SJ 漂移层的 BV（见图 9.17c）在每个 n 型或 p 型中的薄板电荷 Q_s 为

$$Q_s = q N_D W_n = q N_A W_p = \varepsilon_r \varepsilon_o E_{critical} \tag{9.30}$$

式中，W_n 和 W_p 分别为 n 型和 p 型柱的宽度。$R_{drift}A$ 表示为

$$R_{drift}A = \rho_{drift} t_{drift}(P/W_n) \tag{9.31}$$

其中沟槽深度由 t_{drift} 近似。由式（2.16）可得，

$$\rho_{drift} = 1/(q N_D \mu_n) \tag{9.32}$$

将式（9.30）和式（9.32）代入式（9.31）会得到以下表达式

$$R_{drift}A = t_{drift} P/(\mu_n Q_s) \tag{9.33}$$

如果假设沿着沟槽的电场是均匀的，

$$t_{drift} = BV/E_{critical\_SJ} \tag{9.34}$$

式中，$E_{critical\_SJ}$ 为 SJ 漂移层的临界电场。由式（9.30）、式（9.33）和式（9.34）可以得出

$$R_{drift}A = BV P/(\varepsilon_r \varepsilon_o \mu_n E_{critical\_SJ}^2) \tag{9.35}$$

由式（2.22）推导得出常规 n 型漂移层的 $R_{drift}A$ 为

$$R_{drift}A = 4BV^2/(\varepsilon_r \varepsilon_o \mu_n E_{critical\_conv}^3) \tag{9.36}$$

式中，$E_{critical\_conv}$ 为常规 n 型漂移层的临界电场。将式（9.35）与式（9.36）进行比较可以清楚地说明 SJ 漂移层的以下优点：

1）$R_{drift}A$ 与 BV 线性增加，这与常规 n 型漂移层的二次增加相反。

2）$R_{drift}A$ 随 P 线性减小，因为根据式（9.30），N_D 和 N_A 必须随 P 的减小而

增大。N_D 越大 ρ_{drift} 也就越小，$R_{drift}A$ 也就越小。

但是请注意，$E_{critical\_SJ}$ 小于 $E_{critical\_conv}$，因为高电场会延伸更大的距离，从而产生增强的碰撞电离[93]。

据作者所知，尚未有报道 GaN SJ MISFET。迄今为止报道的唯一的 4H-SiC SJ MISFET 是具有离子注入形成的漂移层的 V 形槽 MISFET[94]。外延生长的 n 型漂移层厚度为 6.0μm，掺杂浓度为 $3×10^{15} cm^{-3}$，深度为 3μm 的 p 型柱为 $3×10^{16} cm^{-3}$ 高能量（最高 9MeV）铝离子注入形成了箱形轮廓。然后在 SJ 结构上生长 2μm 厚的外延层，并制作了 V 形槽 MISFET 结构（见图 9.23c）。最终的 V_{th} 为 3.4V，与常规 4H-SiC V 形槽 MISFET 的电平相同，其 BV 和 $R_{on}A$ 分别为 820V 和 $0.97mΩ·cm^2$。

对于更高 BV 的 MISFET，SJ 结构的影响应更加清晰。此外，这种 SJ MISFET 的制造将需要沟槽填充外延（如 6.6 节所述）

9.8 总结

本章首先回顾了 MIS 电容器和 AlGaN/GaN 异质结构的物理原理，以提供对 MISFET 和 HFET 结构的理解。关于 $R_{on}A$ 的详细分析以文献报道的 GaN MIS HFET 和 4H-SiC 平面 MISFET 为例进行。本章还介绍了报道的 GaN MESFET、GaN p+栅极 HFET、GaN 沟槽 MISFET、4H-SiC 沟槽 MISFET 和 4H-SiC V 形槽 MISFET 的器件性能。最后，关于 SJ MISFET，本章介绍了电荷平衡的概念，并介绍了最近报道的 4H-SiC V 形槽 SJ MISFET 的器件性能。

参考文献

[1] Nishizawa, J., T. Terasaki, and J. Shibata, "Field Effect Transistor Versus Analog Transistor (Static Induction Transistor)," *IEEE Transactions on Electron Devices*, Vol. 22, No. 4, 1975, pp. 185–197.

[2] Friedrichs, P., et al., "Static and Dynamic Characteristics of 4H-SiC JFETs Designed for Different Blocking Categories," *Materials Science Forum*, Vol. 338–342, 2000, pp. 1243–1246.

[3] Okada, M., et al., "Novel vertical Heterojunction Field-Effect Transistors with Regrown AlGaN/GaN Two-Dimensional Electron Gas Channels on GaN Substrates," *Applied Physics Express*, Vol. 3, No. 5, 2010, pp. 054201-1–054201-3.

[4] Kimoto, T., and J. A. Cooper, *Fundamentals of Silicon Carbide Technology*, Singapore: John Wiley & Sons, 2014, p. 249 and pp. 312–315.

[5] Muller, R. S., and T. I. Kamins, *Device Electronics for Integrated Circuits* (Second Edition), Singapore: John Wiley & Sons, 1986, pp.398–402.

[6] Ambacher, O., et al., "Two-Dimensional Electron Gases Induced by Spontaneous and Piezoelectric Polarization Charges in Ni- and Ga-face AlGaN/GaN heterostructures," *Journal of Applied Physics*, Vol. 85, No. 6, 1999, pp. 3222–3233.

[7] Ploog, K. H., and O. Brandt, "Doping of Group III Nitrides," *Journal of Vacuum Science and Technology A*, Vol. 16, No. 3, 1997, pp. 1609–1614.

[8] Ptak, A. J., et al., "Controlled Oxygen Doping of GaN Using Plasma Assisted Molecular-Beam Epitaxy," *Applied Physics Letters*, Vol. 79, No. 17, 2001, pp. 2740–2742.

[9] Handa, H., et al., "High-Speed Switching and Current-Collapse-Free Operation by GaN Gate Injection Transistors with Thick GaN Buffer on Bulk GaN Substrates," *International Electron Devices Meeting*, San Francisco, Dec. 5–7, 2016, pp. 256–259.

[10] Monemar, B., et al., "Recombination of Free and Bound Excitons in GaN," *Physica Status Solidi B*, Vol. 245, No. 9, 2008, pp. 1723–1740.

[11] Bougrov, V., et al., *Properties of Advanced Semiconductor Materials GaN, AlN, InN, BN, SiC, SiGe* (eds. Levinshtein, M. E., Rumyantsev, S. L., and Shur, M. S.), New York: John Wiley & Sons, 2001, pp. 1–30.

[12] Robertson, J., and B. Falabretti, "Band Offsets of High K Gate Oxides on III-V Semiconductors," *Journal of Applied Physics*, Vol. 100, 2006, pp. 014111-1–014111-8.

[13] Mönch, W., "Elementary Calculation of the Branch-Point Energy in the Continuum of Interface-Induced Gap States," *Applied Surface Physics*, Vol. 117–118, 1997, pp. 380–387.

[14] Baliga, B. J., *Gallium Nitride and Silicon Carbide Power Devices*, Singapore: World Scientific, 2017, p. 305.

[15] Hu, X., et al., "Si_3N_4/AlGaN/GaN Metal-Insulator-Semiconductor Heterostructure Field-Effect Transistors," *Applied Physics Letters*, Vol. 79, No. 17, 2001, pp. 2832–2834.

[16] Hashizume, T., S. Ootomo, and H. Hasegawa, "Suppression of Current Collapse in Insulated Gate AlGaN/GaN Heterostructure Field-Effect Transistors Using Ultrathin Al_2O_3 Dielectric," *Applied Physics Letters*, Vol. 83, No. 14, 2003, pp. 2952–2954.

[17] Kim, D. H., et al., "ALD Al_2O_3 Passivated MBE-Grown AlGaN/GaN HEMTs on 6H-SiC," *Electronics Letters*, Vol. 43, No. 2, 2007, pp. 129–130.

[18] Shi, J., et al., "High Performance AlGaN/GaN Power Switch with HfO_2 Insulation," *Applied Physics Letters*, Vol. 95, No. 4, 2009, pp. 042103-1–042103-3.

[19] Mizue, C., et al., "Capacitance–Voltage Characteristics of Al_2O_3/AlGaN/GaN Structures and State Density Distribution at Al_2O_3/AlGaN/GaN Interface," *Japanese Journal of Applied Physics*, Vol. 50, No. 2, 2011, pp. 021001-1–021001-7.

[20] Mishra, K., et al., "Localization of Oxygen Donor States in Gallium Nitride from First-Principles Calculations," *Physics Review B*, Vol. 76, No. 3, 2011,

pp. 035127-1–035127-9.

[21] Edwards, A. P., et al., "Improved Reliability of AlGaN–GaN HEMTs Using an NH$_3$ Plasma Treatment Prior to SiN Passivation," *IEEE Electron Device Letters*, Vol. 26, No. 4, 2005, pp. 225–227.

[22] Alekseev, E., A. Eisenbach, and D. Pavlidis, "Low Interface State Density AlN/GaN MISFETs," *Electronics Letters*, Vol. 35, No. 24, 1999, pp. 2145–2146.

[23] Selvaraj, S. L., et al., "AlN/AlGaN/GaN Metal-Insulator-Semiconductor High-Electron-Mobility Transistor on 4 in. Silicon Substrate for High Breakdown Characteristics," *Applied Physics Letters*, Vol. 90, No. 17, 2007, pp. 173506-1–173506-3.

[24] Kim, K.-H., N.-W. Kwak, and S. H. Lee, "Fabrication and Properties of AlN Film on GaN Substrate by Using Remote Plasma Atomic Layer Deposition Method," *Electronic Materials Letters*, Vol. 5, No. 2, 2009, pp. 83–86.

[25] Kanechika, M., et al., "A Vertical Insulated Gate AlGaN/GaN Heterojunction Field-Effect Transistor," *Japanese Journal of Applied Physics*, Vol. 46, No. 20–24, 2007, pp. L503–L505.

[26] Nitta, S., et al., "In-Plane GaN/AlGaN Heterostructure Fabricated by Selective Mass Transport Planar Technology," *Materials Science and Engineering B*, Vol. 93, No. 1–3, 2002, pp. 139–142.

[27] Nakamura, S., et al., "Thermal Annealing Effects on p-Type Mg-Doped GaN Films," *Japanese Journal of Applied Physics*, Vol. 31, No. 2B, 1992, pp. L139–L142.

[28] Hurni, C. A., et al., "p-n Junctions on Ga-face GaN Grown by NH$_3$ Molecular Beam Epitaxy with Low Ideality Factors and Low Reverse Currents," *Applied Physics Letters*, Vol. 97, 2010, pp. 222113-1–222113-3.

[29] Yeluri, R., et al., "Design, Fabrication, and Performance Analysis of GaN Vertical Electron Transistors with a Buried p/n Junction," *Applied Physics Letters*, Vol. 106, 2015, pp. 183502-1–183502-5.

[30] Baliga, B. J., *Gallium Nitride and Silicon Carbide Power Devices*, Singapore: World Scientific, 2017, p. 294 and p. 393.

[31] Sze, S. M., and K. K. Ng, *Physics of Semiconductor Devices (Third Edition)*, New Jersey: John Wiley & Sons, 2007, p. 81.

[32] Barker, Jr., A. S., and M. Ilegems, "Infrared Lattice Vibrations and Free-Electron Dispersion in GaN," *Physical Review B*, Vol. 7, No. 2, 1973, pp. 743–750.

[33] Yoshida, T., et al., "Preparation of 3-inch Freestanding GaN Substrates by Hydride Vapor Phase Epitaxy with Void Assisted Separation," *Physica Status Solidi A*, Vol. 205, No. 5, 2008, pp. 1053–1055.

[34] Chowdhury, S., et al., "CAVET on Bulk GaN Substrates Achieved with MBE-grown AlGaN/GaN Layers to Suppress Dispersion," *IEEE Electron Device Letters*, Vol. 33, No. 1, 2012, pp. 61–63.

[35] Nie, H., et al., "1.5-kV and 2.2-mΩ·cm^2 Vertical GaN Transistors on Bulk-GaN Substrates," *IEEE Electron Device Letters*, Vol. 35, No. 9, 2014, pp. 939–941.

[36] Shibata, D., et al., "1.7 kV/1.0 mΩcm^2 Normally-Off Vertical GaN Transistor on GaN Substrate with Regrown p-GaN/AlGaN/GaN Gate Structure," *International Electron Devices Meeting*, San Francisco, Dec. 5–7, 2016, pp. 248–251.

[37] Zhao, J., et al., "3.6 mΩcm², 1726-V 4H-SiC Normally-Off Trenched-and-Implanted Vertical JFETs," *International Symposium on Power Semiconductor Devices and ICs*, Cambridge, April 14–17, 2003, pp. 50–53.

[38] Zhao, J., et al., "Fabrication and Characterization of 11-kV Normally-Off 4H-SiC Trenched-and-Implanted Vertical Junction FETs," *IEEE Electron Device Letters*, Vol. 25, No. 7, 2004, pp. 474–476.

[39] Ishikawa, T., et al., "SiC Power Devices for HEV/EV and a Novel SiC Vertical JFET," *International Electron Devices Meeting*, San Francisco, Dec. 15–17, 2014, pp. 24–27.

[40] Tanaka, Y., et al., "700-V 1.0-mΩcm² Buried Gate SiC-SIT (SiC-BGSIT)," *IEEE Electron Device Letters*, Vol. 27, No. 11, 2006, pp. 908–910.

[41] Baliga, B. J., "The Future of Power Semiconductor Device Technology," *Proceedings of IEEE*, Vol. 89, No. 6, 2001, pp. 822–832.

[42] Shimizu, H., et al., "Static and Switching Characteristics of 3.3 kV Double Channel-Doped SiC Vertical Junction Field Effect Transistor in Cascode Configuration," *Japanese Journal of Applied Physics*, Vol. 54, 2015, pp. 04DP15-1–04DP15-5.

[43] Ueda, D., H. Takagi, and G. Kano, "A New Vertical Power MOSFET Structure with Extremely Reduced On-Resistance," *IEEE Transactions on Electron Devices*, Vol. 32, No. 1, 1985, pp. 2–6.

[44] Fujihira, T., "Theory of Semiconductor Superjunction Devices," *Japanese Journal of Applied Physics*, Vol. 36, No. 10, 1997, pp. 6254–6262.

[45] Shirota, S., and S. Kaneda, "New Type of Varactor Diode Consisting of Multilayer p-n Junctions," *Journal of Applied Physics*, Vol. 49, No. 12, 1978, pp. 6012–6019.

[46] Deboy, G., et al., "A New Generation of High Voltage MOSFETs Breaks the Limit Line of Silicon," *International Electron Devices Meeting*, San Francisco, Dec. 6–9, 1998, pp. 683–685.

[47] Shenoy, J. N., J. A. Cooper, and M. R. Melloch, "High Voltage Double-Implanted Power MOSFETs in 6H-SiC," *IEEE Electron Device Letters*, Vol. 18, No. 3, 1997, pp. 93–95.

[48] Ryu, S.-H., et al., "Design and Process Issues for Silicon Carbide Power DiMOSFETs," *Materials Research Society Symposium Proceedings*, Vol. 640, 2001, pp. H4.5.1–H4.5.6.

[49] Ryu, S.-H., et al., "Large-Area (3.3 mm × 3.3 mm) Power MOSFETs in 4H-SiC," *Materials Science Forum*, Vol. 389–393, 2002, pp. 1195–1198.

[50] Ryu, S.-H., et al., "Development of 10 kV 4H-SiC Power DMOSFETs," *Materials Science Forum*, Vol. 457–460, 2004, pp. 1385–1388.

[51] Tega, N., et al., "Novel Trench-Etched Double-Diffused SiC MOS (TED MOS) for Overcoming Tradeoff Between $R_{on}A$ and Q_{gd}," *International Symposium on Power Semiconductor Devices and ICs*, Kowloon, May 10–14, 2015, pp. 81–84.

[52] Yang, X. Y., B. M. Lee, and V. Misra, "High Mobility 4H-SiC MOSFETs Using Lanthanum Silicate Interface Engineering and ALD Deposited SiO_2," *Materials Science Forum*, Vol. 778–780, 2014, pp. 557–561.

[53] Das, M. K., "Recent Advances in (0001) 4H-SiC MOS Device Technology," *Materials Science Forum*, Vol. 457–460, 2004, pp. 1375–1280.

[54] Fiorenza, P., et al., "SiO$_2$/4H-SiC Interface Doping During Post-Deposition Annealing of the Oxide in N$_2$O and POCl$_3$," *Applied Physics Letters*, Vol. 103, 2013, pp. 153508-1–153508-4.

[55] Okamoto, D., et al., "Improved Inversion Channel Mobility in 4H-SiC MOSFETs on Si Face Utilizing Phosphorus-Doped Oxide," *IEEE Electron Device Letters*, Vol. 31, No. 7, 2010, pp. 710–712.

[56] Lichtenwalner, D. J., et al., "High-Mobility SiC MOSFETs with Chemically Modified Interfaces," *Materials Science Forum*, Vol. 821–823, 2015, pp. 749–752.

[57] Lichtenwalner, D. J., et al., "High-Mobility SiC MOSFETs with Alkaline Earth Interface Passivation," *Materials Science Forum*, Vol. 858, 2016, pp. 671–676.

[58] Modic, A., et al., "High Channel Mobility 4H-SiC MOSFETs by Antimony Counterdoping," *IEEE Electron Device Letters*, Vol. 35, No. 9, 2014, pp. 894–896.

[59] Okamoto, D., et al., "Improved Channel Mobility in 4H-SiC MOSFETs by Boron Passivation," *IEEE Electron Device Letters*, Vol. 35, No. 12, 2014, pp. 1176–1178.

[60] Isaacs-Smith, T., et al., "Boron Passivation for Improved Channel Mobility in 4H-SiC MOSFETs," *MRS Spring Meeting & Exhibit*, Phoenix, Mar. 28–April 1, 2016, paper EP2.1 04.

[61] Sze, S. M., and K. K. Ng, *Physics of Semiconductor Devices (Third Edition)*, New Jersey: John Wiley & Sons, 2007, p. 28 and p. 328.

[62] Noguchi, M., et al., "Determination of Intrinsic Phonon-Limited Mobility and Carrier Transport Property Extraction of 4H-SiC MOSFETs," *International Electron Devices Meeting*, San Francisco, Dec. 4–6, 2017, pp. 9.3.1–9.3.4.

[63] Saha, A., and J. A. Cooper, "A 1-kV 4H-SiC Power DMOSFET Optimized for Low On-Resistance," *IEEE Transactions on Electron Devices*, Vol. 54, No. 10, 2007, pp. 2786–2791.

[64] Wang, S. R., and J. A. Cooper, "Double-Self-Aligned Short-Channel Power DMOSFETs in 4H-SiC," *Device Research Conference*, Pennsylvania, June 22–24, 2009, pp. 277–278.

[65] http://www.wolfspeed.com/materials/products/sic-substrates.

[66] Soler, V., et al., "High Voltage 4H-SiC Power MOSFETs with Boron Doped Gate Oxide," *IEEE Transactions on Industrial Electronics*, Vol. 64, No. 11, 2017, pp. 8962–8970.

[67] Palmour, J. W., et al., "4H-Silicon Carbide Power Switching Devices," *Institute of Physics Conference Series*, Vol. 142, 1996, pp. 813–816.

[68] Otake, H., et al., "Vertical GaN-Based Trench Gate Metal Oxide Semiconductor Field-Effect Transistors on GaN Bulk Substrates," *Applied Physics Express*, Vol. 1, No. 1, 2008, pp. 011105-1–011105-3.

[69] Kodama, M., et al., "GaN-Based Trench Gate Metal Oxide Semiconductor Field-Effect Transistor Fabricated with Novel Wet Etching," *Applied Physics*

Express, Vol. 1, No. 2, 2008, pp. 021104-1–021104-3.

[70] Oka, T., et al., "Vertical GaN-Based Trench Metal Oxide Semiconductor Field-Effect Transistors on a Freestanding GaN Substrate with Blocking Voltage of 1.6 kV," *Applied Physics Express*, Vol. 7, No. 2, 2014, pp. 021022-1–021022-3.

[71] Oka, T., et al., "1.8 mΩcm^2 Vertical GaN-based Trench Metal Oxide Semiconductor Field-Effect Transistors on a Freestanding GaN Substrate for 1.2 kV-Class Operation," *Applied Physics Express*, Vol. 8, No. 5, 2015, pp. 054101-1–054101-3.

[72] Oka, T., et al., "Over 10A Operation with Switching Characteristics of 1.2 kV-class Vertical GaN-Based Trench MOSFETs on a Bulk GaN Substrate," *International Symposium on Power Semiconductor Devices and ICs*, Prague, June 12–16, 2016, pp. 459–462.

[73] Gupta, C., et al., "In Situ Oxide, GaN Interlayer-Based Vertical Trench MOSFET (OG-FET) on Bulk GaN Substrates," *IEEE Electron Device Letters*, Vol. 38, No. 3, 2017, pp. 353–355.

[74] Gupta, C., et al., "Demonstrating >1.4 kV OG-FET Performance with a Novel Double Field-Plated Geometry and the Successful Scaling of Large-Area Devices," *International Electron Devices Meeting*, San Francisco, Dec. 4–6, 2017, pp. 9.4.1–9.4.4.

[75] Sun, M., et al., "High-Performance GaN Vertical Fin Power Transistors on Bulk GaN Substrates," *IEEE Electron Device Letters*, Vol. 38, No. 4, 2017, pp. 509–512.

[76] Zhang, Y., et al., "1200V GaN Vertical Fin Power Field-Effect Transistors," *International Electron Devices Meeting*, San Francisco, Dec. 4–6, 2017, pp. 9.2.1–9.2.4.

[77] Agarwal, A. K., et al., "1.1 kV 4H-SiC Power UMOSFETs," *IEEE Electron Device Letters*, Vol. 18, No. 12, 1997, pp. 586–588.

[78] Sugawara, Y., and K. Asano, "1.4 kV 4H-SiC UMOSFET with Low Specific on Resistance," *International Symposium on Power Semiconductor Devices and ICs*, Kyoto, June, 3–6, 1998, pp. 119–122.

[79] Baliga, B. J., *Fundamentals of Power Semiconductor Devices*, New York: Springer Science, 2008, p. 296.

[80] http://www.ioffe.ru/SVA/NSM/Semicond/Si/electric.html#Ionization.

[81] Ozbek, A. M., and B. J. Baliga, "Planar Nearly Ideal Edge-Termination Technique for GaN Devices," *IEEE Electron Device Letters*, Vol. 32, No. 3, 2011, pp. 300–302.

[82] Niwa, H, J. Suda, and T. Kimoto, "Impact Ionization Coefficients in 4H-SiC Toward Ultrahigh-Voltage Power Devices," *IEEE Transactions on Electron Devices*, Vol. 62, No. 10, 2015, pp. 3326–3333.

[83] Barker, Jr., A. S., and M. Ilegems, "Infrared Lattice Vibrations and Free-Electron Dispersion in GaN," *Physical Review B*, Vol. 7, No. 2, 1973, pp. 743–750.

[84] Patrick, L., and W. J. Choyke, "Static Dielectric Constant of SiC," *Physical Review B*, Vol. 2, No.6, 1970, pp. 2255–2256.

[85] Sze, S. M., and K. K. Ng, *Physics of Semiconductor Devices (Third Edition)*, New Jersey: John Wiley & Sons, 2007, p. 791.

[86] Li, Y., J. A. Cooper, and M. A. Capano, "High Voltage (3 kV) UMOSFETs in 4H-SiC," *IEEE Transactions on Electron Devices*, Vol. 49, No. 6, 2002, pp. 972–975.

[87] Sui, Y., T. Tsuji, and J. A. Cooper, "On-State Characteristics of SiC Power UMOSFETs on 115 μm Drift Layers," *IEEE Electron Device Letters*, Vol. 26, No. 4, 2005, pp. 255–257.

[88] Nakamura, T., et al., "High Performance SiC Trench Devices with Ultra-low Ron," *International Electron Devices Meeting*, Washington, DC, June 5–7, 2011, pp. 599–601.

[89] Harada, S., et al., "Determination of Optimum Structure of 4H-SiC Trench MOSFET," *International Symposium on Power Semiconductor Devices and ICs*, Bruges, June 3–7, 2012, pp. 253–256.

[90] Wada, K., et al., "Fast Switching 4H-SiC V-Groove Trench MOSFETs with Buried p+ Structure," *International Symposium on Power Semiconductor Devices and ICs*, Waikoloa, June 15–19, 2014, pp. 225–228.

[91] Mikamura, Y., et al., "Novel Designed SiC Devices for High Power and High Efficiency Systems," *IEEE Transactions on Electron Devices*, Vol. 62, No. 2, 2015, pp. 382–389.

[92] Uchida, K., et al., "The Optimized Design and Characterization of 1,200 V/2.0 mΩcm^2 4H-SiC V-Groove Trench MOSFETs," *International Symposium on Power Semiconductor Devices and ICs*, Kowloon, May 10–14, 2015, pp. 85–88.

[93] Baliga, B. J., *Silicon Carbide Power Devices*, Singapore: World Scientific, 2005, p. 369.

[94] Masuda, T., R. Kosugi, and T. Hiyoshi, "0.97 mΩcm^2/820 V 4H-SiC Super Junction V-Groove Trench MOSFET," *Materials Science Forum*, Vol. 897, 2016, pp. 483–488.

第 10 章

双极功率二极管和功率开关器件

10.1 引言

如 8.1 节所述，只要漂移层上的压降可忽略不计，单极功率二极管的正向电压降 V_F[在正向电流 I_F 处取值（见图 1.8）]要小于双极功率二极管的 V_F。反之，当额定电压较高时，双极功率二极管的 V_F 则小于单极功率二极管的 V_F。对于 4H-SiC 功率二极管而言，考虑功率耗散的情况下，当反向电压超过 2 kV，工作频率低于 30kHz 时，理论上应首选双极功率二极管[1]。另一方面，基于 1.6kV GaN 双极功率二极管的存储电荷仅是 Si 双极功率二极管的 5.6% 的事实，对于 GaN 双极功率二极管而言，反向电压和工作频率都将降低[2]。这种存储电荷上的差异是由 GaN 较短的载流子寿命引起的。然而，光子回收可以在大注入条件下增加 GaN 的有效载流子寿命（参见第 4 章），从而可以实现具有高的击穿电压（比如 5.0kV）的 GaN 双极功率二极管[3]。双极功率二极管包括 p-n 结二极管（见图 10.1a）和 p-i-n 二极管（见图 10.1b）。此外，p-n-p-n 二极管（即肖克利二极管）（见图 10.1c）也已被制造出来，尽管它在功能上是两端口功率开关器件。所有这些二极管均采用 3.7 节所述的非自对准台面结构。

目前尚未见到有关 GaN 双极功率开关器件的报道。虽然已经可以在独立的 GaN 衬底上制备出 GaN/InGaN 异质结双极晶体管，但是该晶体管是针对 RF 小信号应用而设计的。基于第一个双极功率开关器件是 Si 基双极晶体管（BJT）的事实，本章将简要介绍可能成为第一个 GaN 双极功率开关器件的 n-p-n GaN BJT 的特性。

作为 4H-SiC 功率开关器件家族的重要组成部分，BJT（见图 10.2a）、晶闸管（见图 10.2b）和绝缘栅双极晶体管（IGBT）（见图 10.2c）已经被研制出来了。击穿电压高达 1.7 kV 的 4H-SiC n-p-n BJT 已实现商业化，其特性模型和仿真也已具备[5]。

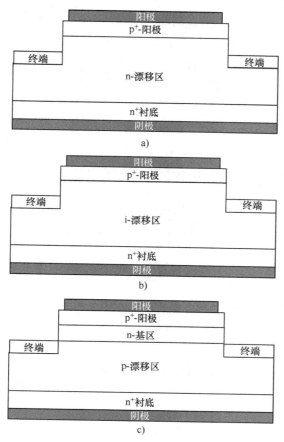

图 10.1　双极功率二极管的横截面示意图：a）p-n 结二极管，b）p-i-n 二极管，以及 c）肖克利二极管（p-n-p-n 二极管）。虽然"i"表示本征，但是通常在宽禁带半导体 p-i-n 二极管的 i 漂移区中做 n 型轻掺杂

晶闸管是通过在肖克利二极管中加入栅电极而得到的（见图 10.1c）。这样一来它可以通过开启得到长时间的导通。通常，通过降低阳极电流到维持电流 I_h 以下来关闭传统晶闸管。然而，栅极可关断晶闸管（GTO）和发射极可关断晶闸管（ETO）可以分别由栅极和发射极负电流来关断。业内已实现耐压高达 22kV 的 4H-SiC GTO[6] 和 4H-SiC ETO[7]。在 BJT 和晶闸管中，控制电极（即

BJT 的基极和晶闸管的栅极）与半导体形成欧姆接触。IGBT 中则使用了绝缘栅结构。IGBT 的名称即源自绝缘栅场效应晶体管和双极晶体管。N 沟道 IGBT（见图 10.2c）的结构与 N 沟道 MISFET（见图 9.17a）相似，只是将 MISFET 中的 n^+ 衬底替换成 IGBT 中的 p^+ 衬底。虽然 IGBT 的背面电极通常被称为集电极，但它的作用其实是 p-n-p BJT 的发射极（见图 10.3）。现已可见耐压高达 27kV 的 4H-SiC N 沟道 IGBT 的报告[8]。

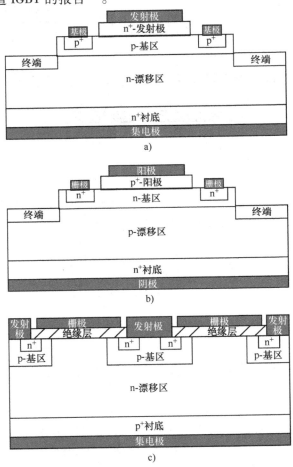

图 10.2 双极功率开关器件的横截面示意图：a) BJT，b) 晶闸管，以及 c) IGBT

本章仅介绍上述平面功率二极管和功率开关器件。但是请悉知，类似于 9.6 节中描述的 MISFET，沟槽和 V 形槽结构也已应用于 IGBT 中[9-11]。

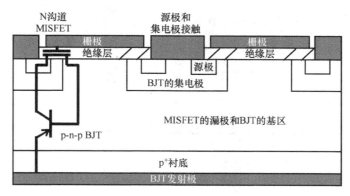

图 10.3 由 n 沟道 MISFET 和 p-n-p BJT 组成的 N 沟道 IGBT 的横截面示意图

10.2 一维 p-n 结二极管的优化设计

第 3.5 节通过式（3.31）推导了当耗尽层宽度 W_{Dn} 小于或等于漂移区厚度 W_n 时，一维 p^+n 型二极管在反偏条件下的耗尽层宽度，用 V_R 替代 $\Psi_{bi}-V$ 可得到：

$$W_{Dn} \approx (2\varepsilon_r\varepsilon_o V_R/qN_D)^{0.5} \tag{10.1}$$

式中，ε_r 为平行于 C 轴的相对介电常数；ε_o 为真空介电常数；V_R 为反向偏压；N_D 为均匀掺杂漂移层的净施主杂质浓度。

在额定电压下，当 $W_n = W_{Dn}$（见图 10.4）且 $W_{Dn} \gg W_{Dp}$ 时，击穿电压 BV 由下式给出：

$$BV \approx E_{max} W_n/2 \tag{10.2}$$

将 GaN 的 ε_r 设为 10.4[12]，将 4H-SiC 的 ε_r 设为 10.0[13]（见 2.7 节），并将 GaN 中的临界击穿电场 3.75MV/cm[14] 和 4H-SiC 中的临界击穿电场 2.5MV/cm[15]（见 2.6 节）分别代入式（10.1）和式（10.2）中，可得到一维 p-n 结二极管在均匀掺杂漂移层下的优化设计（见图 10.5）。

TMG、氨（NH_3）和硅烷（SiH_4）常被用作制备 n 型 GaN 的 MOCVD 气体源（见 6.2 节）。已知在高 NH_3 分压下进行 MOCVD 可以将残留的碳最小化。比如，在 GaN（0001）外延生长过程中，当 NH_3/TMG 流量比超过 1000 时可降低非预期碳受主的浓度（从 $5\times10^{16} cm^{-3}$ 到 $1\sim 2\times10^{16} cm^{-3}$）[16]（见图 10.6 中的空心符号）。假设最小可控 N_D 为 $1.5\times10^{16} cm^{-3}$，可以从图 10.5a 中得到，W_n 为 15μm 时的最高 BV 约为 3kV。以下 BV 的值接近经验测试值：$N_D = 6\times10^{15} cm^{-3}$、$W_n = 40$μm 时 $BV = 3.7$ kV[16]；$N_D = 2\times10^{16} cm^{-3}$、$W_n = 10$μm 时 $BV = 1.1$kV[17]。

10.3节将介绍如何采用非均匀n型掺杂的漂移层来实现更高的BV[3,18,19]。

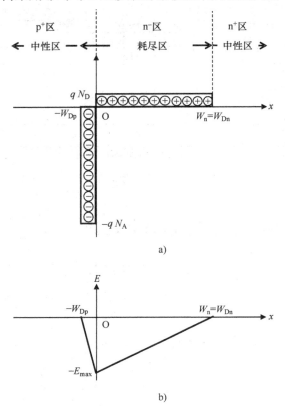

图10.4 当反向$p^+/n^-/n^+$突变结中耗尽区的宽度等于漂移层厚度时：
a）耗尽近似下的空间电荷分布和b）电场分布

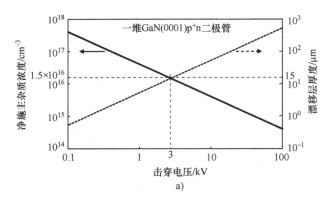

图10.5 耗尽区宽度等于漂移区厚度的突变$p^+/n^-/n^+$结 a）GaN（0001）二极管和b）4H-SiC（0001）二极管的最优化设计

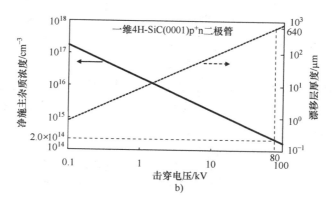

图 10.5 耗尽区宽度等于漂移区厚度的突变 $p^+/n^-/n^+$ 结 a) GaN (0001) 二极管和 b) 4H-SiC (0001) 二极管的最优化设计（续）

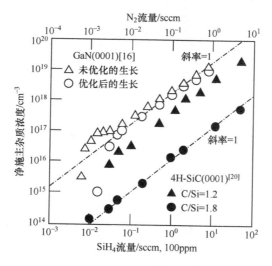

图 10.6 文献[16, 20]中报道的 GaN (0001) 和 4H-SiC (0001) 的外延生长过程中净施主杂质浓度与 SiH_4 和 N_2 流量的关系

另一方面，在 4H-SiC 的 CVD 过程中，可以通过增加输入气体的 C/Si 比来显著降低最小可控 N_D。比如，C/Si 比从 1.2 增加到 1.8，N_D 可以从 $1×10^{16}$ cm^{-3} 降到 $2×10^{14}$ cm^{-3}（见图 10.6 中的实心符号）。根据图 10.5b 和图 10.6 可以预估，当漂移层厚度为 640μm 时，BV 最高可达 80kV。但是，这么厚的漂移层也将大幅度增加 $R_{on}A$。通过减小 W_n 使其小于 W_{Dn}（形成图 10.7a 中所示的 p-i-n 结），电场线围绕的图形将变成梯形，如图 10.7b 中的实线所示。随着漂移层浓度（N_{D1}

的减小，电场线围绕的图形将接近矩形，如图 10.7b 中的虚线所示。这样可以趋近最大 BV，其计算式为

$$BV_{max} = E_{max}/W_n \tag{10.3}$$

4H-SiC 双极二极管常采用 p-i-n 结构，以利用更低以及更可控的 N_D 的优点，10.4 节将对此进行介绍。请注意，这里的"i"尽管表示本征，但通常使用轻掺杂的 n 型层作为 i 漂移层。

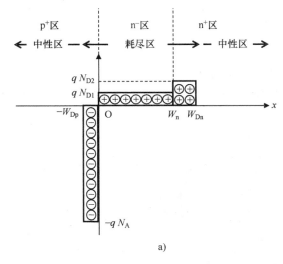

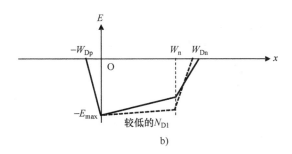

图 10.7 反向偏置的突变 $p^+/n^-/n^+$ 结：a) 耗尽近似下的空间电荷分布和 b) 电场分布，b) 中的虚线表示 N_{D1} 较低时的电场分布

10.3 具有非均匀掺杂漂移层的 GaN p-n 结二极管

为了降低 p-n 结处的峰值电场，研究人员制备了一种具有双漂移层结构

的 GaN p-n 结二极管,其击穿电压超过 3kV。双漂移区结构由掺杂浓度为 1×10^{15} cm^{-3} 的 5μm 厚 n 型 GaN 层以及掺杂浓度为 1×10^{16} cm^{-3} 的 15μm 厚 n 型 GaN 层组成[18]。三漂移区结构也被应用于 BV = 3.48kV 的 GaN p-n 结二极管[19]:一个掺杂浓度为 1×10^{15} cm^{-3} 的 6μm 厚 GaN 层与一个掺杂浓度为 3×10^{15} cm^{-3} 的 11μm 厚 n 型 GaN 层以及一个掺杂浓度为 1.2×10^{16} cm^{-3} 的 15μm 厚 n 型 GaN 层叠加;应用于 BV=5kV 的 GaN p-n 结二极管[3]:一个残余硅浓度小于 2×10^{15} cm^{-3} 的 5.5μm 厚本征 GaN 层与一个硅掺杂浓度为 9×10^{15} cm^{-3} 的 22μm 厚 n 型 GaN 层以及一个硅掺杂浓度为 1.6×10^{16} cm^{-3} 的 5.5μm 厚 n 型 GaN 层叠加。

得益于非本征光子再循环(EPR)(见 4.6 节)已报道的 GaN p-n 结二极管[3,17-19]的差分特征导通电阻 $R_{on}A^{diff}$(见 1.3.2 节)与 4H-SiC p-i-n 二极管[21,22](见图 10.8)相当。但是,由于半导体封装功耗的限制(300 W/cm^2[23]),EPR 在 GaN p-n 结二极管中不起作用。如图 10.9 中的例子所示,5kV GaN p-n 结二极管的最大允许电流密度是 90A/cm^2[3],对应的正向电压是 3.4V。这个正向电压与 3.5kV 4H-SiC JBS 二极管相当(见 8.9 节)(3.1V)[24]。与之相反的是,EPR 在零偏(见图 1.7a)GaN n-p-n 双极晶体管中可以有效工作,10.5.4 节将对此进行介绍。

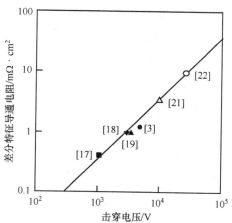

图 10.8 已报道的 GaN p-n 结二极管(实心符号)[3,17-19]和 4H-SiC p-i-n 二极管(空心符号)[21,22])的差分特征导通电阻与击穿电压的关系

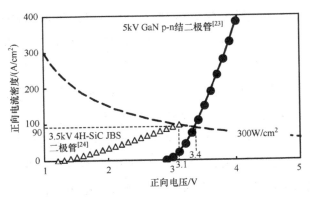

图 10.9 已报道的 5kV GaN p-n 结二极管[3]和 3.5kV 4H-SiC JBS 二极管[24]的电流/电压特性,以及半导体封装功率损耗限制的曲线（300W/cm²）[23]

10.4 4H-SiC p-i-n 二极管

10.4.1 已报道的 4H-SiC p-i-n 二极管结果

在 4H-SiC p-i-n 二极管中采用外延生长的阳极,可以实现 6.2kV 的 BV[25]。据报道,在使用离子注入形成阳极的情况下,器件的正向特性对结深度和激活过程非常敏感[26]。报道中展示了一个 100μm 厚且掺杂浓度为 $1 \sim 3 \times 10^{14}$ cm^{-3} 漂移层的 4H-SiC p-i-n 二极管,其在 100A/cm² 的条件下 V_F = 7.1V,BV 为 8.6kV[27]。铝、碳和硼的共注入技术也被用于形成 4H-SiC p-i-n 二极管的阳极,该二极管具有 40μm 厚,掺杂浓度为 1×10^{15} cm^{-3} 的漂移层,[28]在 100A/cm² 的条件下 V_F = 4.7V,BV 约为 4.5kV。就半导体封装的功耗限制而言,这些 V_F 值是不足的（见图 10.9）。

因为使用铝离子注入不能完全实现电导调制（见 3.7 节）[29],载流子寿命通过外延生长[30]和消除碳空位缺陷（例如碳离子注入[31]和热氧化[32]）得到了提高。由于晶体生长工艺（例如减少了微管和缺陷密度）和外延工艺（例如减少了外延缺陷）得到了改善,因此 4H-SiC 的载流子寿命也得以提高。在 2004 年报道了在 100 A/cm² 的条件下具有 3.9V V_F 的 10kV p-i-n 二极管[30]。它的 i 漂移层厚度为 100μm,掺杂浓度为 2×10^{14} cm^{-3}。在 2012 年报道了使用碳离子注入或者热氧化制作的 4H-SiC p-i-n 二极管[33],在 100 A/cm² 的条件下具有 4.0V 的 V_F。它的 i 漂移层厚度为 120μm,掺杂浓度为 7×10^{13} cm^{-3},BV 经计算为 18.5kV。文献 [21] 和 [22] 分别报道了在 12.9kV BV 下具有 3.3mΩ·cm² $R_{on}A^{diff}$

（漂移区的掺杂浓度为$3\times10^{14}\ cm^{-3}$，厚度为$100\mu m$）以及在26.9kVBV下具有$9.72m\Omega\cdot cm^2 R_{on}A^{diff}$（漂移层的掺杂浓度为$1\sim 2\times10^{14}\ cm^{-3}$且厚度为$268\mu m$）的器件，如图10.8中的空心符号所示。

10.4.2 正偏4H-SiC p-i-n 二极管的存储电荷

在大注入下（见3.6.2节），正向电流密度可通过下式计算：

$$J_F = J_S^{HL}\exp(qV/2kT) \tag{10.4a}$$

$$J_S^{HL} = \{2qD_a N_c^{0.5} N_V^{0.5}\tanh(d/L_a)/L_a/[1-0.25\tanh^4(W_n/2L_a)]^{0.5}\}$$
$$\times\exp[-(E_g+qV_M)/(2kT)] \tag{10.4b}$$

式中，V_M为落在 i 漂移层的电压降；D_a为双极扩散系数，由下式决定：

$$D_a = 2D_n D_p/(D_n+D_p) \tag{10.4c}$$

式中，D_n和D_p分别为电子和空穴的扩散系数；L_a为双极扩散长度，由下式决定：

$$L_a = (D_a\tau_{HL})^{0.5} \tag{10.4d}$$

式中，τ_{HL}为高注入寿命。平均载流子浓度和 i 漂移层的存储电荷可分别表示为

$$n_{ave} = J_F\tau_{HL}/qW_n \tag{10.5}$$

以及

$$Q_s = qW_n n_{ave} = J_F\tau_{HL} \tag{10.6}$$

由式（10.6）可以得到，当J_F固定时，Q_s随着τ_{HL}的增加而增加。因此尽管较高的τ_{HL}可以获得较低的V_F（见10.4.1节），但是有必要通过降低τ_{HL}来减小Q_s。

这里以文献［22］报道的 26.9kV 4H-SiC p-i-n 二极管的正向电流密度/电压特性为例计算Q_s。当半导体封装的功耗限制为 300 W/cm²时[23]，J_F应该小于 70 A/cm²（见图10.10）。若式（10.6）中的τ_{HL}为$3\mu s$[22]，则最大Q_s为2.1×10^{-4}C/cm²。

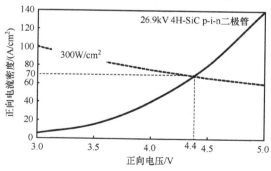

图10.10　已报道的 26.9kV 4H-SiC p-i-n 二极管的电流/电压特性[22]，以及半导体封装功耗限制曲线（300W/cm²）

10.4.3 4H-SiC p-i-n 二极管的反向恢复

如 1.2 节所述，功率二极管在感性负载系统中（如电动机）做反向导通器件使用。由于需要移除导通期间存储在 i 漂移层中的电荷，因此 p-i-n 二极管从导通状态到截止状态的功耗大于从截止状态到导通状态的功耗。在 p-i-n 二极管承受高电压之前产生较大反向电流的现象称为"反向恢复"。

如图 10.11 所示，阳极电流密度从 J_F 以 $-k$ 的斜率减小，直到 W_{Dn} 变为 W_n。在电流密度达到最小值之后（即 $J_R = -kt_{rr}$），电压迅速上升到电源电压[34]。在反向恢复期间被移除的电荷量为

$$Q_{rr} = J_R t_{rr}/2 \quad (10.7)$$

式中，t_{rr} 为反向恢复时间，如图 10.11 所示。

由式 (10.6) 和式 (10.7)，t_{rr} 和 J_R 可以分别表示为

$$t_{rr} = (2\tau_{HL}J_F/k)^{0.5} \quad (10.8)$$

以及

$$J_R = (2k\tau_{HL}J_F)^{0.5} \quad (10.9)$$

例如，当 $\tau_{HL} = 3\mu s$[22]，$J_F = 70 A/cm^2$，$k = 2 \times 10^7 A/cm^2/s$ 时，t_{rr} 和 J_R 分别为 $4.6\mu s$ 和 $92 A/cm^2$。

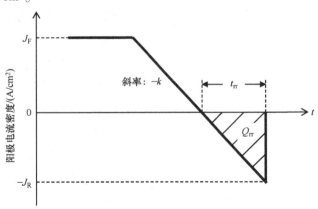

图 10.11 反向恢复期间 p-i-n 二极管的阳极电流密度随时间波形示意图

10.5 n-p-n 双极晶体管

1977 年报道了第一款 SiC BJT（双极晶体管），该器件由掺杂浓度为

10^{20}cm^{-3} 的 n 型发射极层、0.8μm 厚且掺杂浓度为 $4\times10^{17}\text{cm}^{-3}$ 的 p 型基极层和掺杂浓度为 $5\times10^{18}\text{cm}^{-3}$ 的 n 型集电极层组成[35]。然而，由于集电极层掺杂浓度较高，该器件的基极开路集电极发射极击穿电压 BV_{CEO} 相当低（50V）。考虑到该劣势，本节首先从集电极层和基极层的设计角度出发，详细探究造成 BJT 电压降突然降低的"二次击穿"现象的临界集电极电流密度，最后，本节将探讨 GaN BJT 的预期表现以及现有 4H-SiC BJT 的性能。

10.5.1 集电极层设计

本节不考虑电导调制的影响。发射极开路集电极-基极击穿电压 BV_{CBO} 与 p-i-n 二极管的 BV 值相同（见 10.2 节），可以表示为

$$BV_{CBO} = E_{critical}W_N - qN_{D1}W_N^2/(2\varepsilon_r\varepsilon_o) \tag{10.10}$$

漂移区特征导通电阻 $R_{drift}A$ 可以表示为

$$R_{drift}A = W_N/(q\mu_n N_{D1}) \tag{10.11}$$

式中，μ_n 为电子迁移率。

结合式（10.10）和式（10.11），消除 N_{D1}，可以得到下式：

$$R_{drift}A = W_N^3/[2\mu_n\varepsilon_1\varepsilon_o(E_{critical}W_N - BV_{CBO})] \tag{10.12}$$

将式（10.12）关于 x 微分（见图 10.7），可得：

$$dR_{drift}A/dx = W_N^2(2E_{critical}W_N - 3BV_{CBO})/[2\mu_n\varepsilon_1\varepsilon_o(E_{critical}W_N - BV_{CBO})^2] \tag{10.13}$$

当 $dR_{drift}A/dx = 0$ 时，$R_{drift}A$ 达到最小值，即

$$W_{N\_optimum} = (3/2)(BV_{CBO}/E_{critical}) \tag{10.14a}$$

同时有

$$N_{D1\_optimum} = (4/9)(\varepsilon_r\varepsilon_o E_{critical}^2/qBV_{CBO}) \tag{10.14b}$$

BV_{CEO} 与 BV_{CBO} 的关系可以表示为

$$BV_{CEO} = BV_{CBO}(1+\beta_0)^{-1/n} \tag{10.15}$$

式中，β_0 为低电流时共发射极电流增益系数；SiC 的 n 大约为 10[36]。如图 10.12 所示，只要 β_0 大于 10^3，BV_{CEO}/BV_{CBO} 计算值大于 0.5。因此，假设 $BV_{CBO} = 2BV_{CEO}$。式（10.14a）和式（10.14b）可以分别表示为

$$W_{N\_optimum} = 3BV_{CEO}/E_{critical} \tag{10.16a}$$

和

$$N_{D1\_optimum} = (2/9)(\varepsilon_r\varepsilon_o E_{critical}^2/qBV_{CEO}) \tag{10.16b}$$

结果如图 10.13 所示。

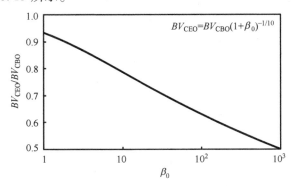

图 10.12 根据式（10.15）计算得到的 BV_{CEO} 与 BV_{CBO} 之比和低电流下共发射极电流增益的关系。此处假设 n 为 10[36]

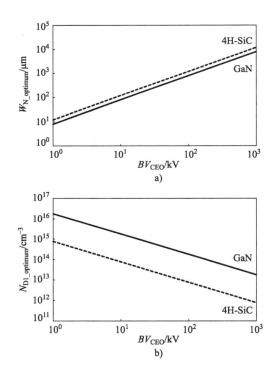

图 10.13 当 $BV_{CEO} = 2BV_{CBO}$ 时，利用式（10.16a）和式（10.16b）计算得到的 a) $W_{N\_optimum}$ 和 b) $N_{D1\_optimum}$ 与 BV_{CEO} 的关系

10.5.2 基极层设计

由式（10.16a）和式（10.16b）确定最佳集电极层的电荷浓度为

$$Q_{\text{collector}} = (2/3)(\varepsilon_r \varepsilon_o E_{\text{critical}}/q) \quad (10.17)$$

可以得到，GaN 的电荷浓度为 $1.4 \times 10^{13} \text{cm}^{-2}$，4H-SiC 的电荷浓度为 $9.2 \times 10^{12} \text{cm}^{-2}$。为防止发射极-集电极穿通，基极电荷浓度应该高于 $Q_{\text{collector}}$。此外，基极层厚度 W_B 应该小于电子扩散长度 L_N（见 3.6.11 节）以获得较高 β_0。

另外，为减小 p 型基极层中因基极电流引起的压降而导致发射极电流过度集中的情况，可以采用较高的基极层掺杂浓度[37]。然而，基极层掺杂浓度应当足够低以保证发射极注入效率 γ_E，其中注入效率为电子从发射极注入基极所产生的发射极电流的占比，表示为[37]

$$\gamma_E = D_{nB} t_E N_{DE}^+ / (D_{nB} t_E N_{DE}^+ + D_{pE} W_B N_{AB}^-) \quad (10.18)$$

式中，D_{nB} 和 D_{pE} 分别为基极层电子扩散系数和发射极层空穴扩散系数；N_{DE}^+ 和 N_{AB}^- 分别为发射极层电离受主浓度和基极层电离施主浓度；t_E 为发射极层厚度。

10.5.3 二次击穿的临界集电极电流密度

对于硅基 BJT，有一种称为二次击穿的现象，该现象限制了器件的安全工作区（即器件正常工作允许的电压电流范围）[38]。二次击穿的临界集电极电流密度 J_{CSB} 表示为[38]

$$J_{CSB} = \varepsilon_r \varepsilon_o v_S E_{\text{critical}}^2 / (2BV_{CBO}) \quad (10.19)$$

式中，v_S 为电子饱和速率（Si 为 $1 \times 10^7 \text{cm/s}$[39]；GaN 为 $2.5 \times 10^7 \text{cm/s}$[40]；4H-SiC 为 $3.3 \times 10^6 \text{cm/s}$[41]）。当 $BV_{CBO} = 1.2\text{kV}$（即 $BV_{CEO} \approx 0.6\text{kV}$），计算可得，Si 基 BJT 的 J_{CSB} 为 0.17kA/cm^2，GaN 基 BJT 的 J_{CSB} 为 120kA/cm^2，4H-SiC 基 BJT 的 J_{CSB} 为 15kA/cm^2。对于 Si 基 BJT，J_{CSB} 低于工作电流密度导致了二次击穿。对于 GaN 基和 4H-SiC 基 BJT，J_{CSB} 大于工作电流密度，换而言之，GaN 基和 4H-SiC 基 BJT 不会发生二次击穿。

10.5.4 GaN BJT

虽然 GaN n-p-n BJT 在蓝宝石衬底上形成了准纵向结构，但实验还是证明了该器件具有零偏电流/电压特性[42]。对于零偏的 BJT，有望实现本征光子循环（EPR）。如 4.6 节所述，p 型 GaN 中 EPR 的横向延伸距 p 型电极边缘约 $10\mu m$。

因此，间距20μm基极插值的GaN BJT具有最佳面积利用率$R_{on}A$。4.6节确定了最有效的受主能级E_A^{eff}，基于此假设，文献［43］中仿真了具有非自对准基台结构的0.6kV量级GaN BJT（$BV_{CBO} = 1.2$kV）的电流/电压特性。GaN BJT的$R_{on}A$可以低于目前最先进的0.6kV量级4H-SiC沟槽MISFET的$R_{on}A$[44]。然而，使得基极电流密度最小的BJT结构仍有待优化。

10.5.5 SiC BJT

第一款SiC BJT（报道于2001年）具有1.8kV的BV_{CEO}和20的$β_0$[45]。随后，文献［46］中报道的2.7kV BJT的$β_0$提升至50，文献［47］中将21kV BJT的$β_0$提升至63，文献［48］中将0.6kV BJT的$β_0$提升至73，文献［49］中将0.27kV BJT的$β_0$提升至110，文献［50］中将0.95kV BJT的$β_0$提升至134。2011年，通过在4H-SiC（0001）和（000$\bar{1}$）衬底上制作的BJT，实现了257和335的$β_0$值，尽管并没有报道该BJT的BV_{CEO}[51]。这些$β_0$值是通过优化的器件几何形状和发射极-基极结的连续外延生长，并结合深能级缺陷减少工艺来提高基极层的载流子寿命来实现的。

10.6 肖克利二极管

由于还没有GaN或者4H-SiC肖克利二极管被报道过，本节将介绍硅基肖克利二极管的基本特性，作为理解晶闸管特性的基础。一维肖克利二极管的基本结构如图10.14所示。该结构包含了三个p-n结：J_1、J_2和J_3。p_2掺杂区域的宽度（W_{p2}）远大于其他三个区域的宽度，并且p_2的掺杂量级远低于其他三个区域。

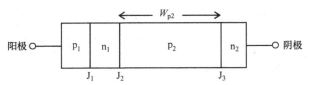

图10.14 一维肖克利二极管的结构示意图。在典型的硅基肖克利二极管中，净掺杂浓度是$p_1 \approx 10^{20}$cm^{-3}，$n_1 \approx 10^{18}$cm^{-3}，$p_2 \approx 10^{14}$cm^{-3}，$n_2 \approx 10^{19}$cm^{-3}

10.6.1 肖克利二极管的反向阻断

在反向阻断模式下，J_1 和 J_3 是反向偏置的，而 J_2 是正向偏置的。绝大部分反向偏压被 p_2 区域所承受。若在击穿时耗尽区宽度小于 W_{p2}，则雪崩倍增将导致在 BV_R 下发生击穿（见图 10.15）；反之，穿通（即 J_2 与 J_3 短路）将导致了在 BV_R 下发生击穿。

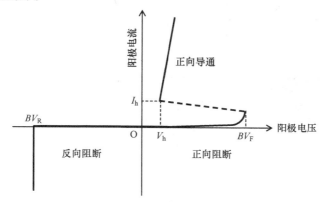

图 10.15 肖克利二极管的电流/电压特性示意图

10.6.2 肖克利二极管的正向阻断

在正向阻断模式下，J_1 和 J_3 是正向偏置的，而 J_2 是反向偏置的。因此绝大部分施加的阳极电压由 J_2 承受。当阳极电压达到 BV_F 时（见图 10.15），有电子-空穴对产生（见 2.6 节所述）（见图 10.16a 中的（1））。产生的电子和空穴分别注入 n_1 和 p_2 中性区中（见图 10.16a 中的（2）和（2'）），导致 n_1 和 p_2 耗尽宽度的减小（见图 10.16b 中的（3）和（3'））。为了保持电中性，p_1 和 n_2 的耗尽区宽度同时被减小（见图 10.16b 中的（4）和（4'）），从而增大了 J_1 和 J_2 的正向偏压（见图 10.16b 中的（5）和（5'））。这一增大的正向偏压提高了从 p_1 中性区到 n_1 中性区的空穴注入（见图 10.16b 中的（6）），以及从 n_2 中性区到 p_2 中性区的电子注入（见图 10.16b 中的（6'）），从而进一步提高了 p_1 和 n_2 中性区的电子与空穴的注入（图 10.16a 中的（2）和（2'））。这一正反馈将阳极电压从 BV_F 减小到 V_h（见图 10.15）并且实现了正向导通。

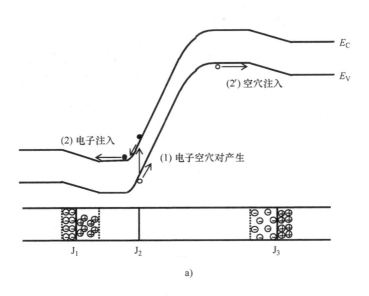

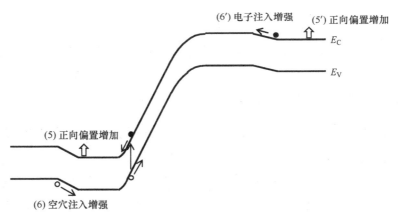

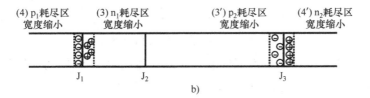

图 10.16 在图 10.15 中阳极电压达到 BV_F a) 之前 b) 之后的能带示意图

10.7　SiC 晶闸管

将栅电极连接到肖克利二极管的 n_1 区域（见图 10.17）就得到了 SiC 晶闸管（见图 10.2b）。如图 10.18 所示，晶闸管甚至可以在阳极电压低于 BV_F 的情况下（见图 10.15）由栅电极注入电子开启。2014 年报道的一个 22kV 4H-SiC GTO，采用了 160μm 厚，掺杂浓度为 $2\times10^{14}cm^{-3}$ 的 p 型漂移区，在 25A/cm^2 条件下获得了 5V 的 V_F[6]。然而，由于 10.8 节中介绍的高压 SiC IGBT 取得了成功，因此，SiC 晶闸管目前没有得到充分的发展。

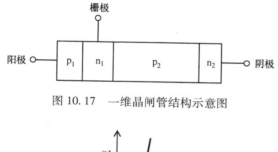

图 10.17　一维晶闸管结构示意图

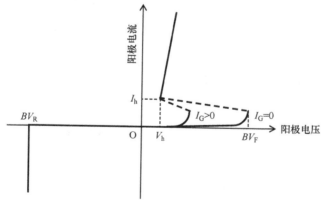

图 10.18　晶闸管的电流/电压特性示意图

10.8　SiC 绝缘栅型晶闸管

自从 2007 年报道了一个 4H-SiC n 型沟道的 IGBT 有能力达到 13kV 的阻断电压后[52]，10kV[53]、12.5kV[54]、16kV[55]、22kV[56] 和 27kV[8] n 型沟道 IGBT 相继被报道。由于电导调制效应的存在（见 3.7 节），IGBT 的 $R_{dirft}A$ 远低于 MI-

SFET（见 9.7 节）。相反，IGBT 的开关损耗远大于 MISFET，其原因与 10.4.4 节中描述的 p-i-n 二极管的反向恢复类似。由于 SiC IGBT 的基础工作原理在他处已经进行了详细描述[57,58]，本节仅简要介绍 4H-SiC IGBT 的关断特性。

为了关断 IGBT，栅极偏置需要减小到零电位（见图 10.19a）。当栅极偏置低于阈值电压时，沟道中的电子电流停止。但是，在感性负载条件下，集电极电流被存储在 n 型漂移区中的空穴所维持，如图 10.19b 所示。另一方面，当栅极电压低于阈值电压时，集电极偏压立即增加（见图 10.19c）。

在 4H-SiC 外延生长时进行可控钒掺杂已经被证实可以减少空穴的注入；文献 [59] 中报道了由氮和钒混合掺杂的外延层拥有低于 20ns 的少数载流子寿命。

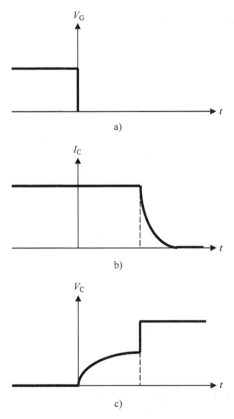

图 10.19　驱动电感负载的 n 沟道 IGBT 的关断波形：
a) 栅极偏置，b) 集电极电流和 c) 集电极偏置

10.9 总结

本章首先介绍了一维结构的优化设计,使读者对 GaN p-n 结二极管有了清晰的了解。然后,为了清楚认识 4H-SiC p-i-n 二极管,本章介绍了存储电荷和反向恢复效应。关于功率开关器件,本章首先从集电极/基极设计和二次击穿的角度讨论了 n-p-n BJT。随后,为了清楚地了解 4H-SiC 晶闸管,本章还介绍了肖克利二极管。最后,本章简要介绍了已被报道的 4H-SiC n 沟道 IGBT 的阻断电压及其典型的关断波形。

参 考 文 献

[1] Morisette, D. T., and J. A. Cooper, Jr., "Theoretical Comparison of SiC PiN and Schottky Diodes Based on Power Dissipation Considerations," *IEEE Transactions on Electron Devices*, Vol. 49, No. 9, 2002, pp. 1657–1664.

[2] Kajitani, R., et al., "A High Current Operation in a 1.6 kV GaN-Based Trenched Junction Barrier Schottky Diode," *Solid State Devices and Materials*, Sapporo, Sept. 27–30, 2015, pp. 1056–1057.

[3] Ohta, H., et al., "5.0 kV Breakdown-Voltage Vertical GaN p-n Junction Diodes," *Extended Abstracts of International Solid State Devices and Materials*, Sendai, Sep. 19–22, 20017, pp. 671–672.

[4] Lochner, Z., et al., "NpN-GaN/In$_x$Ga$_{1-x}$N/GaN Heterojunction Bipolar Transistor on Freestanding GaN Substrate," *Applied Physics Letters*, Vol. 99, 2011, pp. 193501-1–193501-3.

[5] Brooks, S. E., "Modeling and Simulation of 1700-V 8-A GeneSiC Superjunction Transistor," *Thesis and Dissertation*, University of Arkansas, Fayetteville, Aug. 2016.

[6] Palmour, J. W., "Silicon Carbide Power Device Development for Industrial Markets," *International Electron Devices Meeting*, San Francisco, Dec. 15–17, 2014, pp. 1–8.

[7] Song, X., et al., "22 kV SiC Emitter Turnoff (ETO) thyristor and Its Dynamic Performance Including SOA," *International Symposium on Power Semiconductor Devices and ICs*, Hong Kong, May 10–14, 2015, pp. 277–280.

[8] Brunt, E. V., et al., "27-kV, 20-A 4H-SiC n-IGBTs," *Materials Science Forum*, Vol. 821–823, 2015, pp. 847–850.

[9] Chang, H. R., and B. J. Baliga, "500-V n-Channel Insulated Gate Bipolar Transistor with a Trench Gate Structure," *IEEE Transactions on Electron Devices*, Vol. 36, No. 9, 1989, pp. 1824–1829.

[10] Singh, R., et al., "High Temperature SiC Trench Gate p-IGBT," *IEEE Transactions on Electron Devices*, Vol. 50, No. 3, 2003, pp. 774–784.

[11] Zhang, Q., et al., "10 kV Trench Gate IGBTs on 4H-SiC," *International Symposium on Power Semiconductor Devices and ICs*, Santa Barbara, May 23–26, 2005, pp. 159–162.

[12] Barker, Jr., A. S., and M. Ilegems, "Infrared Lattice Vibrations and Free-Electron Dispersion in GaN," *Physical Review B*, Vol. 7, No. 2, 1973, pp. 743–750.

[13] Patrick, L., and W. J. Choyke, "Static Dielectric Constant of SiC," *Physical Review B*, Vol. 2, No.6, 1970, pp. 2255–2256.

[14] Ozbek, A. M., and B. J. Baliga, "Planar Nearly Ideal Edge-Termination Technique for GaN Devices," *IEEE Electron Device Letters*, Vol. 32, No. 3, 2011, pp. 300–302.

[15] Niwa, H., J. Suda, and T. Kimoto, "Impact Ionization Coefficients in 4H-SiC Toward Ultrahigh-Voltage Power Devices," *IEEE Transactions on Electron Devices*, Vol. 62, No. 10, 2015, pp. 3326–3333.

[16] Kizilyalli, I. C., et al., "Vertical Power p-n Diodes Based on Bulk GaN," *IEEE Transactions on Electron Devices*, Vol. 62, No. 2, 2015, pp. 414–421.

[17] Hatakeyama, Y., et al., "Over 3.0 GW/cm^2 Figure-of-Merit GaN p-n Junction Diodes on Freestanding GaN Substrates," *IEEE Electron Device Letters*, Vol. 32, No. 12, 2011, pp. 1674–1676.

[18] Hatakeyama, Y., et al., "High-Breakdown-Voltage and Low-Specific-on-Resistance GaN p-n Junction Diodes on Freestanding GaN Substrates Fabricated Through Low-Damage Field-Plate Process," *Japanese Journal of Applied Physics*, Vol. 52, 2013, pp. 028007-1–028007-3.

[19] Nomoto, K., et al., "GaN-on-GaN p-n Power Diodes with 3.48 kV and 0.95 mΩcm^2: A Record High Figure-of-Merit of 12.8 GW/cm^2," *International Electron Devices Meeting*, Washington, D.C., Dec. 7–9, 2015, pp. 237–240.

[20] Kimoto, T., and J. A. Cooper, *Fundamentals of Silicon Carbide Technology*, Singapore: John Wiley & Sons, 2014, p. 92.

[21] Sundarsesan, S., et al., "12.9 kV SiC PiN Diodes with Low On-State Drops and High Carrier Lifetimes," *Materials Science Forum*, Vol. 717–720, 2012, pp. 949–952.

[22] Kimoto, T., et al., "Progress in Ultrahigh-Voltage SiC Devices for Future Power Infrastructure," *International Electron Devices Meeting*, San Francisco, Dec. 15–17, 2014, pp. 36–39.

[23] Zhang, Q., et al., "12-kV p-Channel IGBTs with Low On-Resistance in 4H-SiC," *IEEE Electron Device Letters*, Vol. 29, No. 9, 2008, pp. 1027–1029.

[24] Brosselard, P., et al., "High Temperature Behaviour of 3.5 kV 4H-SiC JBS Diodes," *International Symposium on Power Semiconductor Devices and ICs*, Jeju, May 27–30, 2007, pp. 285–288.

[25] Sugawara, Y., et al., "6.2 kV 4H-SiC Pin Diode with Low Forward Voltage Drop," *Materials Science Forum*, Vol. 338–342, 2000, pp. 1371–1374.

[26] Chilukuri, R. K., et al., "High Voltage p-n Junction Diodes in Silicon Carbide Using Field Plate Edge Termination," *MRS Symposium Proceedings*, Vol. 572, 1999, pp. 81–86.

[27] Singh, R., "Silicon Carbide Bipolar Power Devices—Potentials and Limits," *MRS Symposium Proceedings*, Vol. 640, 2001, pp. H4.2.1–H4.2.12.

[28] Fedison, J. B., et al., "Al/C/B Co-implanted High Voltage 4H-SiC Pin Junction Rectifiers," *Materials Science Forum*, Vol. 338–342, 2000, pp. 1367–1370.

[29] Kimoto, T., et al., "Promise and Challenges of High-Voltage SiC Bipolar Power Devices," *Energies*, Vol. 9, No. 11, 2016, pp. 908-1–908-15.

[30] Das, M. K., et al., "High Power, Drift-Free 4H-SiC PiN Diodes," *International Journal of High Speed Electronics and Systems*, Vol. 14, No. 3, 2004, pp. 860–864.

[31] Storasta, L., and H. Tsuchida, "Reduction of Traps and Improvement of Carrier Lifetime in 4H-SiC Epilayers by Ion Implantation," *Applied Physics Letters*, Vol. 90, No. 6, 2007, pp. 062116-1–062116-3.

[32] Hiyoshi, T., and T. Kimoto, "Reduction of Deep Levels and Improvement of Carrier Lifetime in n-Type 4H-SiC by Thermal Oxidation," *Applied Physics Express*, Vol. 2, No. 4, 2009, pp. 041101-1–041101-3.

[33] Nakayama, K., et al., "Characteristics of a 4H-SiC Pin Diode with Carbon Implantation/Thermal Oxidation," *IEEE Transactions on Electron Devices*, Vol. 59, No. 4, 2012, pp. 895–901.

[34] Baliga, B. J., *Gallium Nitride and Silicon Carbide Power Devices*, Singapore: World Scientific, 2017, pp. 203–212.

[35] Von Muench, W., P. Hoeck, and E. Pettenpaul, "Silicon Carbide Field-Effect and Bipolar Transistors," *International Electron Devices Meeting*, Washington, D.C., Dec. 5–7, 1977, pp. 337–339.

[36] Chow, T. P., "High-Voltage SiC and GaN Power Devices," *Microelectronics Engineering*, Vol. 83, No. 1, 2006, pp. 112–122.

[37] Baliga, B. J., *Gallium Nitride and Silicon Carbide Power Devices*, Singapore: World Scientific, 2017, p. 442.

[38] Huang, A. Q., and B. Zhang, "The Future of Bipolar Power Transistors," *IEEE Transactions on Electron Devices*, Vol. 48, No. 11, 2001, pp. 2535–2543.

[39] Jacoboni, C., et al., "A Review of Some Charge Transport Properties of Silicon," *Solid State Electronics*, Vol. 20, No. 2, 1977, pp. 77–89.

[40] Bhapkar, U. V., and M. S. Shur, "Monte Carlo Calculation of Velocity-Field Characteristics of Wurtzite GaN," *Journal of Applied Physics*, Vol. 82, No. 4, 1997, pp. 1649–1655.

[41] Sankin, V. I., and A. A. Lepneva, "Electron Saturated Vertical Velocities in Silicon Carbide Polytypes," *Materials Science Forum*, Vol. 338–342, 2000, pp. 769–771.

[42] Terano, A., T. Tsuchiya, and K. Mochizuki, "Characteristics of GaN-Based Bipolar Transistors on Sapphire Substrates with the n-Type Emitter Region Formed Using Si-ion Implantation," *IEEE Transactions on Electron Devices*, Vol. 61, No. 10, 2014, pp. 3411–3416.

[43] Mochizuki, K., "Vertical GaN Bipolar Devices: Gaining Competitive Advantage from Photon Recycling," *Physica Status Solidi A*, Vol. 214, No. 3, 2017, pp. 1600489-1–1600489-8.

[44] Nakamura, T., et al., "High Performance SiC Trench Devices with Ultra-low R_{on}," *International Electron Devices Meeting*, Washington, D.C., June 5–7, 2011, pp. 599–601.

[45] Ryu, S.-H., et al., "1800V NPN Bipolar Junction Transistors in 4H-SiC," *IEEE Electron Device Letters*, Vol. 22, No. 3, 2001, pp. 124–126.

[46] Ghandi, R., et al., "Fabrication of 2700-V 12 mΩ·cm² Non-ion-implanted 4H-SiC BJTs with Common-Emitter Current Gain of 50," *IEEE Electron Device Letters*, Vol. 29, No. 10, 2008, pp. 1135–1137.

[47] Miyake, H., et al., "21-kV SiC BJTs with Space-Modulated Junction Termination Extension," *IEEE Electron Device Letters*, Vol. 29, No. 11, 2008, pp. 1598–1600.

[48] Miyake, H., T. Kimoto, and J. Suda, "Improvement of Current Gain in 4H-SiC BJTs by Surface Passivation with Deposited Oxides Nitrided in N_2O or NO," *IEEE Electron Device Letters*, Vol. 32, No. 3, 2011, pp. 285–287.

[49] Zhang, Q., et al., "4H-SiC BJTs with Current Gain of 110," *Solid State Electronics*, Vol. 52, No. 7, 2008, pp. 1008–1010.

[50] Nonaka, N., et al., "A New High Current Gain 4H-SiC Bipolar Junction Transistor with Suppressed Surface Recombination Structure: SSR-BJT," *Materials Science Forum*, Vol. 615–617, 2009, pp. 821–824.

[51] Miyake, H., T. Kimoto, and J. Suda, "4H-SiC BJTs with Record Current Gains of 257 on (0001) and 335 on ($000\bar{1}$)," *IEEE Electron Device Letters*, Vol. 32, No. 7, 2011, pp. 841–843.

[52] Das, M. K., et al., "A 13-kV 4H-SiC n-Channel IGBT with Low $R_{diff,on}$ and Fast Switching," *Materials Science Forum*, Vol. 600–603, 2008, pp. 1183–1186.

[53] Zhang, Q., et al., "SiC Power Devices for Microgrids," *IEEE Transactions on Power Electronics*, Vol. 25, No. 12, 2010, pp. 2889–2896.

[54] Ryu, S., et al., "Ultra High Voltage (> 12 kV), High Performance 4H-SiC IGBTs," *International Symposium on Power Semiconductor Devices and ICs*, Bruges, July 3–7, 2012, pp. 257–260.

[55] Yonezawa, Y., et al., "Low V_f and Highly Reliable 16 kV Ultrahigh Voltage SiC Flip-Type n-Channel Implantation and Epitaxial IGBT," *International Electron Devices Meeting*, Washington, D.C., Dec. 9–11, 2013, pp. 164–167.

[56] Brunt, E. V., et al., "22 kV, 1 cm², 4H-SiC n-IGBTs with Improved Conductivity Modulation," *International Symposium on Power Semiconductor Devices and ICs*, Waikoloa, June 15–19, 2014, pp. 358–361.

[57] Kimoto, T., and J. A. Cooper, *Fundamentals of Silicon Carbide Technology*, Singapore: John Wiley & Sons, 2014, pp. 373–392.

[58] Baliga, B. J., *Gallium Nitride and Silicon Carbide Power Devices*, Singapore: World Scientific, 2017, Chapter 17.

[59] Miyazawa, T., et al., "Vanadium Doping in 4H-SiC Epitaxial Growth for Carrier Lifetime Control," *Applied Physics Express*, Vol. 9, 2016, pp. 11130-1– 11130-4.

第 11 章

边 缘 终 端

11.1 引言

当 SBD 没有终端保护时,其周围的电场在肖特基电极的边缘尖锐处(见图 11.1a 中的 A 点)会增强。尽管利用氩离子注入对 GaN[1,2] 和 SiC[3] 表面进行非晶化能够有效缓解这种电场增强效应,但反向漏电仍然较大,因为这可能是残留的晶体缺陷所造成的。在硅 SBD 中通常引入 p^+ 保护环(见图 11.1b)来屏蔽这种电场增强效应。然而,对于 GaN SBD 来说,镁离子注入工艺还不成熟(见 7.3.1 节)。另一方面,对于 4H-SiC SBD 来说,注入的铝离子基本不扩散(见 7.4 节),这会使保护环的外缘(见图 11.1b 中的 B 点)发生明显的电场聚集效应。同样地,若铝离子注入的 4H-SiC p-n 平面结没有终端保护时,其外缘(见图 11.1c 中的 C 点)的电场也会增强。因此,除了采用表面非晶化工艺和 p^+ 保护环,要制造接近理想击穿电压 BV 的 GaN 和 4H-SiC 功率器件,边缘终端结构也是必不可少的。

对于 GaN 功率器件的边缘终端技术,场板(FP)是经常用到的一种结构[4](见图 11.2a),尤其是无 p^+ 保护环[5](见图 11.2 b 和 c)或者带有 p^+ 保护环的台面场板(MFP)[6](见图 11.3)。另一方面,相对于由离子注入形成的终端结构,场限环(FLR)(见图 11.4a)[7] 的优势在于可与 p^+ 注入工艺步骤同时形成。然而,对于高电压等级,由于环间距受到光刻精度的限制,FLR 的优势较小。因此,一种名为终结端拓展(JTE)的结构(见图 11.4b)[8] 更适用于高电压 4H-SiC 功率器件的制造[9]。

JTE 的最大问题是对受主掺杂浓度与铝离子注入剂量 ϕ 比值的激活率有着高敏感性。为了减轻对 ϕ 变化的敏感性,横向变掺杂(VLD)技术被(见图 11.4c

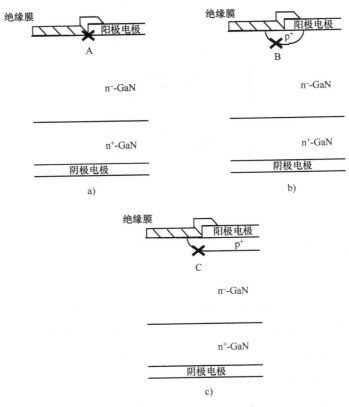

图 11.1 a) 没有终端保护的 SBD，b) 带有保护环终端的 SBD 和
c) 没有终端保护的离子注入 p+n 平面结的横截面示意图

应用于硅功率器件[10]。VLD 结构的制造方法是：p 型杂质离子通过条形掩模注入，掩膜开口面积从 p^+ 阳极区域到边缘区域逐渐减少。随后的 p^+ 推结扩散过程中形成了一个由主结到边缘掺杂浓度逐渐降低、结深逐渐减小的掺杂轮廓。另一方面，对于 4H-SiC 功率器件，由于铝的扩散率可以忽略不计，以及硼的异常扩散，因此无法进行推结式扩散（见 7.4 节）。针对此问题，几种改进的 JTE 结构被提出——包括最小化光刻和离子注入步骤数量的 JTE 结构：空间调制 JTE（SM-JTE）（见图 11.4d）[11]，反掺杂 JTE（CD-JTE）（见图 11.4e）[12]，以及混合 JTE（见图 11.4h）[13]，它们结合了环辅助 JTE（RA-JTE）（见图 11.4f）和多浮空区 JTE（MFZ-JTE）（见图 11.4g）。需要注意的是，4H-SiC 的表面通常用 SiO_2 钝化，SiO_2/4H-SiC 界面的正电荷（Q_f）可以补偿 JTE 区域中负极性的电离受主杂质[14-16]。因此，对 ϕ 有较大容差的 JTE 对 Q_f 也可以具有较大的容差[12]。

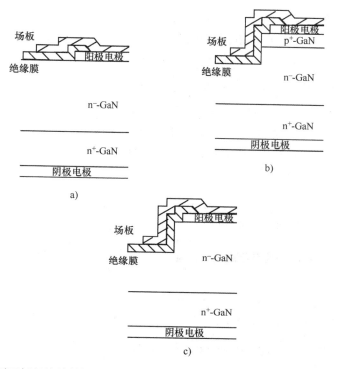

图 11.2　a）平面场板终端结构的 GaN SBD，b）台面场板终端结构的 GaN p-n 结二极管和 c）台面场板终端结构的 GaN SBD 的截面示意图

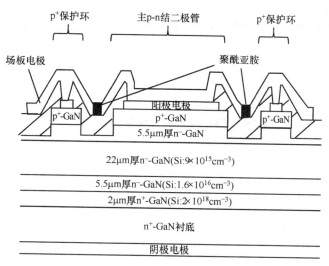

图 11.3　带有保护环的台面场板终端的 GaN p-n 结二极管[6]的横截面示意图

传统的 FP、FLR 和 JTE 终端结构在一些教科书中已经有了详细的描述[17,18]。例如，对于 FP 来说，绝缘介质厚度和 FP 长度对 4H-SiC 肖特基和 p-n 结的 BV 的影响已经可以通过数值计算得到[18]。对于用 FLR 终端的 p-n 结的 BV，在单个场限环的情况下，其解析表达式也已经给出[18]，而数值计算结果在用多个场限环终端的 4H-SiC p-n 结的情况下得到了例证[17]。对于 JTE，数值计算得出的 4H-SiC p-n 结的 BV 可以表示为单区[18]和双区[17] JTE 中受主杂质浓度的函数。因此，本章只介绍用于 GaN 功率器件的 MFP 以及用于 4H-SiC 功率器件的 SM-JTE、CD-JTE 和混合 JTE。

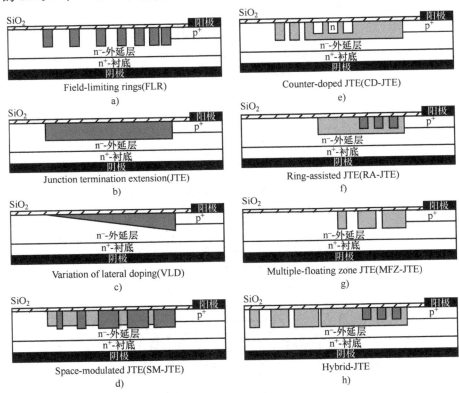

图 11.4 a) FLR，b) JTE，c) 横向变掺杂（VLD），d) 空间调制 JTE（SM-JTE），
e) 反向掺杂 JTE（CD-JTE），f) 环辅助 JTE（RA-JTE），g) 多浮动区 JTE
（MFZ-JTE）和 h) 混合 JTE 的截面示意图

11.2 GaN 功率器件的 MFP

11.2.1 不带保护环的 MFP

对于采用无保护环 MFP 终端结构的 GaN p-n 结二极管（见图 11.2b），绝缘层上的场板电极覆盖到阳极电极边缘，使 n⁻型 GaN 层中的耗尽区沿着台面扩展。继 2011 年这种 MFP 首次应用于 GaN p-n 结二极管后[5]，2014 年和 2015 年分别报道了 GaN MISFET MFP 的终端结构[19]和 GaN SBD MFP 的终端结构[20]（见图 11.2c）。这些器件中使用的 MFP 边缘终端结构是在 SiO_2/旋涂玻璃（SOG）绝缘层[5]或 SiO_2/Al_2O_3 绝缘层[19,20]上形成的。

2012 年，通过仿真研究了具有高 k 电介质 FP 终端结构的 4H-SiC 功率器件[21]。根据该研究结果，与使用相同厚度的 SiO_2 膜层相比，高 k 膜层会产生更宽的耗尽区域。在仿真实验中，Si_3N_4 和 HfO_2 被用作平面 FP 中的绝缘层（见图 11.2a）。2016 年，由 $CeO_2:SiO_2$ 以 2:1 的比例构成的 0.7μm 厚的混合氧化物膜被用作 GaN p-n 结二极管的 MFP[22]。虽然 CeO_2 的相对介电常数 k 为 26，但混合氧化物的 k 降低到了 12.3[22]，与 SiO_2（3.9）相比，它更接近 GaN 的介电常数（10.4[23]）。这种结构产生了更均匀的反向电流和雪崩击穿免疫特性。如图 11.5 中的实心圆所示，使用混合氧化物（$CeO_2:SiO_2$）的 MFP 终端处理的 GaN p-n 结二极管在 2.2 kV 时表现出稳定的击穿特性。相反，使用 SiO_2 的 MFP 终端处理的 GaN 二极管在 2.2 kV 时突然击穿并损害（见图 11.5 中的空心圆）。

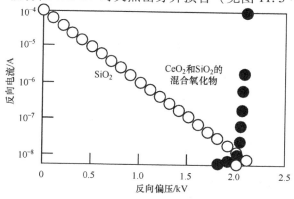

图 11.5 分别使用 SiO_2（空心圆）和 CeO_2 和 SiO_2 的混合氧化物（实心圆）作为介质层的 MFP 终端处理的 GaN p-n 结二极管的反向电流/电压特性[22]

11.2.2 保护环辅助 MFP

图 11.3[6]所示为带有保护环辅助 MFP 终端结构的 GaN p-n 结二极管的横截面示意图。在主 p-n 结二极管外侧引入 p$^+$保护环，并在保护环和主 p-n 结二极管之间插入聚酰亚胺电阻。如果没有这样的电阻，当对 p-n 结二极管施加反向偏压时，它的最外环可能会损坏。该电阻在 p-n 结二极管和保护环之间产生一个电压降，从而将 BV 从 4.8 kV 提高到 5.0 kV[6]。

11.3 用于 4H-SiC 功率器件的 SM-JTE

对于 SM-JTE 终端结构，具有调制宽度和间距的多个环被嵌入均匀掺杂的 JTE 区域中。图 11.6 显示了应用于 4H-SiC p-i-n 二极管（i 层掺杂浓度为 $1\times10^{14}\,\text{cm}^{-3}$，厚度为 $150\mu\text{m}$）的 $600\mu\text{m}$ 宽的 SM-JTE 的击穿电压仿真值[24]。环和 JTE 区域的铝掺杂剂量分别为 $1.8\times10^{13}\,\text{cm}^{-2}$ 和 $4.5\times10^{12}\,\text{cm}^{-2}$。从图 11.6 中可以清楚地看出，与均匀掺杂的 JTE 相比，SM-JTE 提供了更宽的 JTE 最佳掺杂剂量窗口，以获得近乎理想状态的 BV。

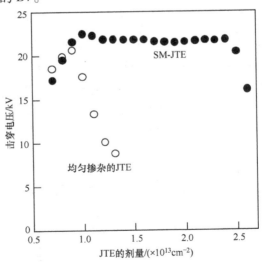

图 11.6 报道的 4H-SiC p-i-n 二极管（具有 $600\mu\text{m}$ 宽的 JTE）的击穿电压随 JTE 剂量变化的仿真结果。i 层的厚度和掺杂浓度分别为 $150\mu\text{m}$ 和 $1\times10^{14}\,\text{cm}^{-3}$。对于 SM-JTE 来说，获得近乎理想击穿电压的最佳掺杂窗口被显著扩大[24]。SM-JTE 已被应用于 21kV 4H-SiC p-i-n 二极管[25]和 21kV 4H-SiC BJT[26]

11.4 用于 4H-SiC 功率器件的 CD-JTE

对于 CD-JTE 终端结构,在均匀的 p 型掺杂的 JTE 区域中使用 n 型反掺杂产生多区域的效果[12]。如图 11.7 中的空心圆所示,当 Q_f 为正时,以均匀掺杂 JTE 为终端的 4H-SiC p-i-n 二极管的 BV 降低。相比之下,用 CD-JTE 作为终端的 p-i-n 二极管的 BV 在 $Q_f = 6 \times 10^{12} \text{ cm}^{-2}$ 以下几乎没有退化(见图 11.7 中的实心圆),这显示出 CD-JTE 终端结构的 BV 对于工艺变化具有出色的鲁棒性[12]。

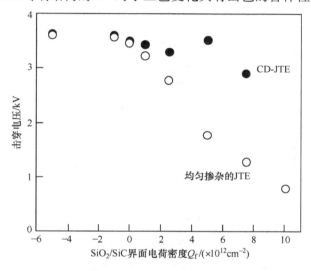

图 11.7 报道中具有 100μm 宽的 CD-JTE(实心圆)和均匀掺杂 JTE(空心圆)的 4H-SiC p-i-n 二极管的模拟击穿电压对 SiO_2/SiC 界面电荷密度的依赖性。i 层的厚度和掺杂浓度分别为 30μm 和 $2 \times 10^{15} \text{cm}^{-3}$ [12]

11.5 用于 4H-SiC 功率器件的混合型 JTE

RA-JTE 把浮动环插入均匀掺杂区域中(见图 11.4f)[9,27],通过控制分散的注入区域(即区域带)的宽度,剂量以 VLD 的方式在 MFZ-JTE 中渐变分布(见图 11.4g)[28]。Sung 和 Baliga 首先对 RA-JTE 和 MFZ-JTE 进行了优化。对于具有 40μm 厚、$2 \times 10^{15} \text{cm}^{-3}$ 掺杂浓度漂移层的 4H-SiC p-i-n 二极管,RA-JTE 和 MFZ-JTE 的优化宽度分别为 120μm 和 90μm[29]。优化后的 RA-JTE 中使用了六个 3μm 宽的 p+ 环，环间距从 3~8μm 不等，第 i 个区域的宽度 W_i（i=2, 3, ⋯, 10）为 10（μm）/[1.07×(i-1)]。将优化过的 RA-JTE 的宽度缩小到 90μm，然后将宽度变窄了的 RA-JTE 放置在 p+ 主结附近，并将 90μm 宽的 MFZ-JTE 放置在 RA-JTE 旁边，构建了一个混合型 JTE（见图 11.3h）。如图 11.8 所示，优化过的 RA-JTE 和均匀掺杂 JTE 的 BV 在高剂量掺杂条件下都会明显降低。这种降低是由于 JTE 外边缘的高电场[29]。由于优化过的 MFZ-JTE 的 BV 对于掺杂剂量的依赖性与优化过的 RA-JTE、均匀掺杂 JTE 的 BV 对于掺杂剂量的依赖性相反，因此对于混合 JTE 来说，获得接近理想击穿电压的最佳剂量窗口被扩大。

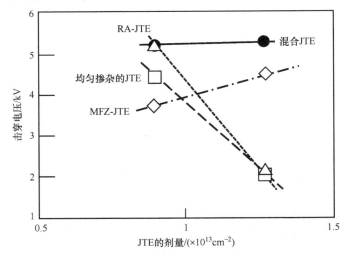

图 11.8 报道中 4H-SiC p-i-n 二极管不同 JET 终端结构的击穿电压（实测值）与 JTE 剂量的相关性，终端结构包括 180μm 宽的混合 JTE、120μm 宽的 RA-JTE、90μm 宽的 MFZ-JTE 和 120μm 宽的均匀掺杂 JTE。i 层的厚度和掺杂浓度分别为 40μm 和 $2×10^{15} cm^{-3}$。假设激活率为 70%，通过 $1.3×10^{13} cm^{-2}$ 和 $1.8×10^{13} cm^{-2}$ 的注入剂量可以估算出 JTE 掺杂剂量为 $9×10^{12} cm^{-2}$ 和 $1.25×10^{13} cm^{-2}$ [29]

11.6 总结

本章介绍了用于 GaN 功率器件的台面场板终端结构，以及用于 4H-SiC 功率器件的空间调制，反向掺杂和混合 JTE 的改进型终端结构。由于相关研究仍在积极进行中，因此读者也可以参考有关垂直型 GaN 和 4H-SiC 功率器件终端结构的最新论文。

参 考 文 献

[1] Ozbeck, A. M., and B. J. Baliga, "Planar Nearly Ideal Edge Termination Technique for GaN Devices," *IEEE Electron Device Letters*, Vol. 32, No. 3, 2011, pp. 300–302.

[2] Ozbeck, A. M., and B. J. Baliga, "Finite-Zone Argon Implant Edge-Termination For High-Voltage GaN Schottky Rectifiers," *IEEE Electron Device Letters*, Vol. 32, No. 10, 2011, pp. 1361–1363.

[3] Alok, D., B. J. Baliga, and P. K. McLarty, "A Simple Edge Termination for Silicon Carbide with Nearly Ideal Breakdown Voltage," *IEEE Electron Device Letters*, Vol. 15, No. 10, 1994, pp. 394–395.

[4] Conti, F., and M. Conti, "Surface Breakdown in Silicon Planar Diodes Equipped with Field Plate," *Solid-State Electronics*, Vol. 15, No. 1, 1972, pp. 93–105.

[5] Nomoto, K., et al., "Over 1.0 kV GaN p-n Junction Diodes on Freestanding GaN Substrates," *Physica Status Solidi A*, Vol. 208, No. 7, 2011, pp. 1535–1537.

[6] Ohta, H., et al., "5.0 kV Breakdown-Voltage Vertical GaN p-n Junction Diodes," *Extended Abstracts of International Solid State Devices and Materials*, Sendai, Sep. 19–22, 20017, pp. 671–672.

[7] Kao, Y. C., and E. D. Wolley, "High-Voltage Planar p-n Junctions," *Proceedings of IEEE*, Vol. 55, No. 8, 1967, pp. 1409–1414.

[8] Temple, V. A. K., and W. Tantraporn, "Junction Termination Extension for Near-Ideal Breakdown Voltage in p-n Junctions," *IEEE Transactions on Electron Devices*, Vol. 33, No. 10, 1986, pp. 1601–1608.

[9] Perez, R., et al., "Planar Edge Termination Design and Technology Considerations for 1.7-kV 4H-SiC PiN Diodes," *IEEE Transactions on Electron Devices*, Vol. 52, No. 10, 2005, pp. 2309–2316.

[10] Stengl, R., and U. Gösele, "Variation of Lateral Doping—A New Concept to Avoid High Voltage Breakdown of Planar Junction," *International Electron Devices Meeting*, Washington, D.C., Dec. 1–4, 1985, pp. 154–157.

[11] Feng, G., J. Suda, and T. Kimoto, "Space-Modulated Junction Termination Extension for Ultrahigh-Voltage p-i-n Diodes in 4H-SiC," *IEEE Transactions on Electron Devices*, Vol. 59, No. 2, 2012, pp. 414–418.

[12] Huang, C., et al., "Counter-doped JTE, an Edge Termination for HV SiC Devices with Increased Tolerance to the Surface Charge," *IEEE Transactions on Electron Devices*, Vol. 62, No. 2, 2015, pp. 354–358.

[13] Sung, W., and B. J. Baliga, "A Comparative Study 4500-V Edge Termination Techniques for SiC Devices," *IEEE Transactions on Electron Devices*, Vol. 64, No. 4, 2017, pp. 1647-1652.

[14] Sheridan, D. C., et al, "Comparison and Optimization of Edge Termination Techniques for SiC Power Devices," *International Symposium on Power Semiconductor Devices and ICs*, Osaka, June 4–7, 2001, pp. 191–194.

[15] Matsushima, H., et al, "Measuring Depletion-Layer Capacitance to Analyze a Decrease in Breakdown Voltage of 4H-SiC Diodes," *Journal of Applied Physics*, Vol. 119, 2016, 154506-1-154506-6.

[16] Matsushima, H, et al, "Analyzing Charge Distribution in the Termination Area of 4H-SiC Diodes by Measuring Depletion-Layer Capacitance," *Japanese Journal of Applied Physics*, Vol. 55, 2016, 04ER17-1-04ER17-5.

[17] Kimoto, T., and J. A. Cooper, *Fundamentals of Silicon Carbide Technology*, Singapore: John Wiley & Sons, 2014, pp. 427–434.

[18] Baliga, B. J., *Gallium Nitride and Silicon Carbide Power Devices*, Singapore: World Scientific, 2017, pp. 90–115.

[19] Oka, T., et al., "Vertical GaN-Based Trench Metal Oxide Semiconductor Field-Effect Transistors on a Freestanding GaN Substrate with Blocking Voltage of 1.6 kV," *Applied Physics Express*, Vol. 7, No. 2, 2014, pp. 021022-1–021022-3.

[20] Tanaka, N. et al., "50A Vertical GaN Schottky Barrier Diode on a Freestanding GaN Substrate with Blocking Voltage of 790V," *Applied Physics Express*, Vol. 8, 2015, pp. 071001-1–071001-3.

[21] Song, Q.-W., et al., "Simulation Study on 4H-SiC Power Devices with High-k Dielectric FP Terminations," *Diamond & Related Materials*, Vol. 22, 2012, pp. 42–47.

[22] Yoshino, M., et al., "High-k Dielectric Passivation for GaN Diode with a Field Plate Termination," *Electronics*, Vol. 5, No. 2, 2016, pp. 15-1–15-7.

[23] Barker, Jr., A. S., and M. Ilegems, "Infrared Lattice Vibrations and Free-Electron Dispersion in GaN," *Physical Review B*, Vol. 7, No. 2, 1973, pp. 743–750.

[24] Kimoto, T., et al., "Progress in Ultrahigh-Voltage SiC Devices for Future Power Infrastructure," *International Electron Devices Meeting*, San Francisco, Dec. 15–17, 2014, pp. 36–39.

[25] Niwa, H., J. Suda, and T. Kimoto, "21.7 kV 4H-SiC PiN Diode with a Space-Modulated Junction Termination Extension," *Applied Physics Express*, Vo. 5, No. 6, 2012, pp. 1598-1600.

[26] Miyake, H., et al., "21 kV SiC BJTs with Space-Modulated Junction Termination Extension," *IEEE Electron Device Letters*, Vol. 33, No. 11, 2012, pp. 1598-1600.

[27] Kinoshita, K., et al., "Guard Ring Assisted RESURF: A New Termination Structure Providing Stable and High Breakdown Voltage for SiC Power Devices," *International Symposium on Power Semiconductor Devices and ICs*, Santa Fe, June 4–7, 2002, pp. 253–256.

[28] Sung, W., et al., "A New Edge Termination Technique for High-Voltage Devices in 4H-SiC-Multiple-Floating-Zone Junction Termination Extension," *IEEE Electron Device Letters*, Vol. 32, No. 7, 2011, pp. 880–882.

[29] Sung, W., and B. J. Baliga, "A Near Ideal Edge Termination Technique for 4500 V 4H-SiC Devices: The Hybrid Junction Termination Extension," *IEEE Electron Device Letters*, Vol. 37, No. 12, 2016, pp. 1609–1612.

第 12 章
垂直型 GaN 和 SiC 功率器件可靠性

12.1 引言

本章将介绍已经报道的垂直型 GaN 和 4H-SiC 功率器件可靠性测试相关内容。基于 JEDEC[1]和 JEITA[2]中的硅功率器件相关标准,可靠性测试的内容包括高温反偏(HTRB)、高温栅偏(HTGB)、高温高湿反偏(H3TRB)、热循环(TC)、高温工作(HTO)和地面宇宙辐照等。

12.2 HTRB 应力耐受性

HTRB 测试验证半导体芯片漏电的长期稳定性。在最高工作环境温度下,通过在半导体芯片或者功率模块(见图 12.1)上施加反向电压(功率器件 80%的额定反向阻断能力)来进行 HTRB 考核。来自于制造过程中沾污的或者焊料中残留的可动离子聚集在高电场区域产生了表面电荷。该表面电荷能够改变功率器件中的电场并引入额外的漏电[3]。

研究表明衬底晶向会显著影响垂直型 GaN 功率器件 HTRB 耐受性。在 GaN 轴向外延生长时,由于螺旋位错周围的螺旋生长会形成大的六边形小丘(见 6.2 节),在引入零点几度的轻微倾斜后会让台阶式生长完全覆盖螺旋生长(见 6.5 节)。用台阶式生长的 231 个(77 个/批次×3 批次)GaN p-n 结二极管(额定电压为 1200V 和 10A)通过了 150℃、960V、1000h 的 HTRB 考核,同时在 788h HTRB 试验前后,GaN 常开结型场效应晶体管的阈值电压 V_{th}、导通电阻 $R_{on}A$ 和正向漏极电流均未变化[4]。对于常关型晶体管，文献 [5] 中对 1.7kV GaN P⁺栅极异质结场效应晶体管的 V_{th} 和关态漏电流 I_{DS} 进行了 300h 以上的稳定性考核研究（漏源偏置 V_{DS} = 400V），如 9.5.3 节（见图 12.2）所示。

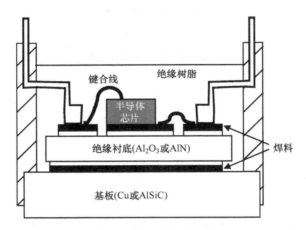

图 12.1　典型功率器件截图

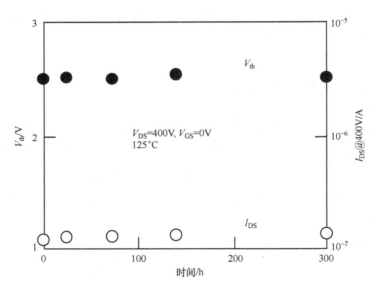

图 12.2　GaN 衬底上垂直型 GaN 晶体管 HTRB 测试结果[5]

MISFET 器件（见 9.7 节）的 HTRB 测试不仅对漏极的 p-n 结施加应力，而且对栅极绝缘层施加应力。对商用 4H-SiC MISFET（额定电压为 1200V 和 42A）在温度 150℃、960V 试验条件下进行考核[6]，378h HTRB 应力后，四个 4H-SiC

MISFET 中一个出现漏源短路，另外三个 MISFET 的漏电流随应力时间不同程度的增加如图 12.3 所示。漏电流退化可能的机理包括漏-源亚阈值漏电、带-带-隧穿主导的 p-n 结漏电以及界面陷阱引起的漏电[6]。

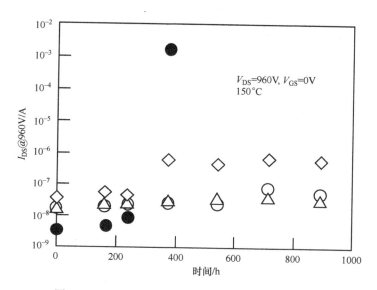

图 12.3 商用 4H-SiC MISFET 的 HTRB 测试结果[6]

此外，在温度为 152℃，额定电压为 90% 的试验条件下，对商用 1.2 ~ 1.7kV 4H-SiC MISFET 功率模块进行 HTRB 测试[7]。五个模块中有三个未通过测试，表明 4H-SiC MISFET 模块对 HTRB 应力的耐受性需要进一步提升。

12.3 HTGB 应力耐受性

HTGB 测试是在高温下给器件长时间施加栅-源偏压 V_{GS}，监测 V_{th} 的变化。对商用的 4H-SiC MISFET（1200V，42A）在 150℃ 下进行考核[6]，如图 12.4 所示。因为 V_{GS} 应力产生的电场使电子隧穿进入或流出绝缘层陷阱[8,9]，加载正向和负向偏压 V_{GS} 导致 V_{th} 正向和负向移动。

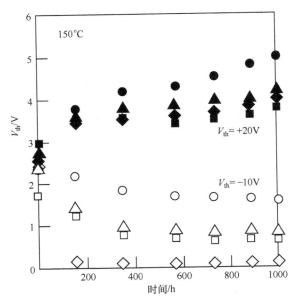

图 12.4 常用 4H-SiC MISFET 的 HTGB 测试结果
（实心：$V_{GS}=+20V$ 和 150℃；空心：$V_{GS}=-10V$ 和 150℃ [6]）

12.4 H3TRB 应力耐受性

在 H3TRB 测试中，在高湿度、高温和高反向偏压下，水分会加速渗透穿过绝缘树脂（见图 12.1），以此来验证功率模块的可靠性。硅 IGBT 功率模块，封闭水膜与芯片有源区的金属层相连，使其成为负电极，进而加剧了腐蚀的形成[10]。GaN p-n 结二极管的 H3TRB 测试（在 85℃、85% 相对湿度和 80V 下）研究表明，850h 后，一些二极管由于芯片边缘进水而导致分层失效。改善封装介电层后，没有二极管在进一步的 H3TRB 测试中失效[4]。

在 90℃、90% 相对湿度和 65% 额定电压下对商用 0.3~0.8kV 4H-SiC MISFET 电源模块进行的 H3TRB 测试结果表明，690h 和 860h 后，五分之二的模块发生失效；前者是由于与 MISFET 芯片相连的二极管失效而引起的，后者是由于栅氧失效引起的[7]。并且三个通过 H3TRB 测试的模块的 $R_{on}A$ 也会增加，这表明 4H-SiC MISFET 功率模块对 H3TRB 应力的耐受性需要进一步提高。

12.5 TC 应力耐受性

TC 测试会在功率模块上引起加速的热机械应力，该功率模块由几种具有不同热膨胀系数的材料组成。对于硅 IGBT 功率模块，键合线剥离是最主要的失效形式[11]。研究表明 77 个/批次×5 批次总数 385 个 GaN p-n 结二极管（额定电压为 1200V 和 10A）时通过了温度从 −65 ~150℃ 的 1000 个周期的 TC 测试[4]。另据报道，当温度从 −50 ~150℃ 变化时，5 个 4H-SiC MISFET 商业功率模块通过了 1186 个循环的 TC 测试[7]。

12.6 HTO 应力耐受性

当激发的少数载流子与有缺陷的电子重组时，重组能量会以非辐射的方式释放，引起缺陷的局部振动，导致缺陷本身重新排列[12,13]。尽管 GaN 器件具有高位错密度，但由于位错运动下较大的剪切应力，其应力寿命依然很长[14]。在正向电流密度为 $20kA/cm^2$，结温为 275℃ 下，对 77 个 GaN p-n 结二极管进行的 HTO 测试结果表明电流/电压特性并未退化[4]。

与之相反，在 4H-SiC 双极器件的漂移层中，BPD 处的电子-空穴复合（见 2.2.3 节）会引起复合增强堆叠层错的成核和扩展，从而增加正向电压 V_F[13,15]。4H-SiC 中的位错运动不是剪切应力而是通过降低材料系统的能量（见 2.2.3 节）引起的[15]。重组引起的堆叠层错也会降低 4H-SiC 单极器件的性能。例如，研究表明正向偏置（$V_{GS}=-10V$）1h，100μm 厚，$6×10^{14}$ cm^{-3} 掺杂漂移层的 4H-SiC MISFET 的体二极管（见图 12.5a），V_F 以和 4H-SiC p-i-n 二极管中观察到的方式增加（见图 12.5b）[16]。另一研究表明正向偏置不同 BPD 密度的 4H-SiC MPS 二极管（见 8.6.1 节），V_F 随注入时间的增加而增加（见图 12.6）[17]。降低 BPD 密度是提高 4H-SiC 功率器件的耐受性的最重要问题。

12.7 地面宇宙辐射耐受性

地球不断地受到高能粒子，主要是质子以及轻、重核[18]的宇宙粒子的辐射作用。这些粒子与空气分子碰撞，形成次级粒子的级联到达地面。尽

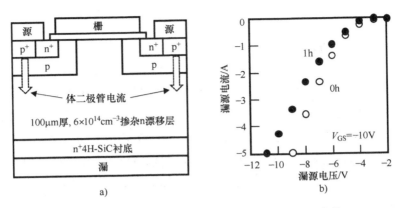

图 12.5 a）横截面示意图，b）4H-SiC MISFET 中体
二极管的电流/电压特性[16]

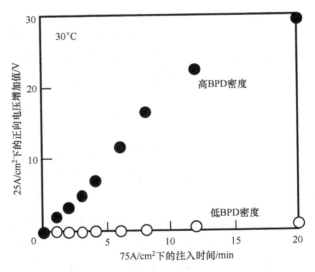

图 12.6 4H-SiC MPS 二极管的正向电压随
注入时间的增加而增加[17]

管这种地面宇宙辐射是由中子、质子、介子、μ 介子，电子和电磁波组成的，但在海平面，中子占了其中的 95%[19]。对于硅功率器件，研究表明中子会与晶格原子相互作用导致失效。产生的反冲原子或分裂产物沿其轨迹产生电荷尖峰。因为电子被吸引到漏极或集电极，而空穴被吸引到源极或栅电极，电子和空穴导致电流瞬变，使寄生双极晶体管导通烧毁器件。栅极下的空穴堆积会增强栅介质周围的电场而导致灾难性失效，从而引发栅

穿故障[20]。

然而，在 GaN 和 4H-SiC 功率器件的情况下，中子诱发的故障还没有很完善的理论理解[20-26]。例如，在一项研究中，GaN p-n 结二极管暴露于 1MeV $3 \times 10^{13} \mathrm{cm}^{-2}$ 的中子中[21]，$R_{on}A$ 从 2~3mΩ·cm² 增加了 3mΩ·cm²，而击穿电压 BV 从 1740V 减小了 80V。$R_{on}A$ 和 BV 的性能退化表明，GaN p-n 结二极管在通态性能和截止态性能都受到漂移区内辐射引起的缺陷的影响。另一方面，比较商用硅和 4H-SiC MISFET 对地面中子辐射的耐受性时[20]，发现尽管中子使 SiC MISFET 的故障少于硅 MISFET，但对于不同的供应商，SiC MISFET 的故障模式也有所不同。这些结果表明，必须进一步研究垂直型 GaN 和 4H-SiC 功率器件对地面宇宙辐射的耐受性。

12.8 总结

本章简要介绍了在垂直型 GaN 和 4H-SiC 功率器件上进行的 HTRB、HTGB、H3TRB、TC、HTO 和地面宇宙辐射测试。相关研究仍在积极进行中，读者可以参考有关此类可靠性测试的最新论文。

参 考 文 献

[1] JEDEC, *Solid State Technology Association, Temperature, Bias, and Operating Life*, JESD22-A108D, 2005.

[2] JEITA, *Environmental and Endurance Test Methods for Semiconductor Devices*, EIAJ ED-4701/100, Tokyo, Japan, 2001.

[3] Lutz, J., et al., *Semiconductor Power Devices Physics, Characterization, Reliability*, Heidelberg: Springer, 2011, pp. 380–411.

[4] Kizilyalli, I. C., et al., "Reliability Studies of Vertical GaN Devices Based on Bulk GaN Substrates," *Microelectronics Reliability*, Vol. 55, No. 9–10, 2015, pp. 1654–1661.

[5] Shibata, D., et al., "1.7 kV/1.0 mΩcm² Normally-Off Vertical GaN Transistor on GaN Substrate with Regrown p-GaN/AlGaN/GaN Gate Structure," *International Electron Devices Meeting*, San Francisco, Dec. 5–7, 2016, pp. 248–251.

[6] Yang, L., and A. Castellazzi, "High Temperature Gate-Bias and Reverse-Bias Tests on SiC MOSFETs," *Microelectronics Reliability*, Vol. 53, 2013, pp. 1771–1773.

[7] Ionita, C., and M. Nawaz, "End User Reliability Assessment of 1.2–1.7 kV Commercial SiC MOSFET Power Modules," *International Reliability Physics Symposium*, Monterey, April 2–6, 2017, pp. WB-1.1–WB-1.6.

[8] Lelis, A., et al., "Time Dependence of Bias-Stress-Induced SiC MOSFET Threshold Voltage Instability Measurements," *IEEE Transactions on Electron Devices*, Vol. 55, No. 8, 2008, pp. 1835–1840.

[9] Lelis, A., et al., "Temperature-Dependence of SiC MOSFET Threshold-Voltage Instability," *Materials Science Forum*, Vol. 600–603, 2009, pp. 807–810.

[10] Minzari, D., et al., "Electrochemical Migration on Electronic Chip Resistors in Chloride Environments," *IEEE Transactions on Device Materials Reliability*, Vol. 9, No. 3, 2009, pp. 392–402.

[11] Wang, H., et al., "Transitioning to Physics-of-Failure as a Reliability Driver in Power Electronics," *IEEE Journal of Emerging and Selected Topics in Power Electronics*, Vol. 2, No. 1, 2014, pp. 97–113.

[12] Kimmering, L. C., "Recombination Enhanced Defect Reactions," *Solid-State Electronics*, Vol. 21, 1978, pp. 1391–1401.

[13] Maeda, K., K. Suzuki, and M. Ichihara, "Recombination Enhanced Dislocation Glide in Silicon Carbide Observed In-Situ by Transmission Electron Microscopy," *Microscopy, Microanalysis, Microstructures*, Vol. 4, 1993, pp. 211–220.

[14] Sugiura, L., "Dislocation Motion in GaN Light-Emitting Devices and Its Effect on Device Lifetime," *Journal of Applied Physics*, Vol. 81, No. 4, 1997, pp. 1633–1638.

[15] Skowronski, M., and S. Ha, "Degradation of Hexagonal Silicon-Carbide-Based Bipolar Devices," *Journal of Applied Physics*, Vol. 99, 2006, pp. 011101-1–011101-24.

[16] Agarwal, A., "A New Degradation Mechanism in High-Voltage SiC Power MOSFETs," *IEEE Electron Devices Letters*, Vol. 28, No. 7, 2007, pp. 587–589.

[17] Caldwell, J. D., "Recombination-Induced Stacking Fault Degradation of 4H-SiC Merged-PiN-Schottky Diodes," *Journal of Applied Physics*, Vol. 106, 2009, pp. 044504-1–044504-6.

[18] Soelkner, G., "Ensuring the Reliability of Power Electronic Devices with Regard to Terrestrial Cosmic Radiation," *Microelectronics Reliability*, Vol. 58, 2016, pp. 39–50.

[19] Ziegler, J. F., "Terrestrial Cosmic Ray Intensities," *IBM Journal of Research and Development*, Vol. 40, No. 1, 1996, pp. 19–39.

[20] Akturk, A., "Single Even Effects in Si and SiC Power MOSFETs due to Terrestrial Neutrons," *IEEE Transactions on Nuclear Science*, Vol. 64, No. 1, 2017, pp. 529–535.

[21] King, M. P., et al., "Performance and Breakdown Characteristics of Irradiated Vertical GaN P-i-N Diodes," *IEEE Transactions on Nuclear Science*, Vol. 62, No. 6, 2015, pp. 2912–2918.

[22] Rashed, K., et al., "Terrestrial Neutron Induced Failure In Silicon Carbide Power MOSFETs," *IEEE Radiation Effects Data Workshop*, Paris, July 14–18, 2014, pp. 1–4.

[23] Akturk, A., R. Wilkins, and J. McGarrity, "Terrestrial Neutron Induced Failures in Commercial SiC Power MOSFETs at 27C and 150C," *IEEE Radiation Effects Data Workshop*, Boston, July 13–17, 2015, pp. 115–119.

[24] Asai, H., et al., "Tolerance Against Terrestrial Neutron-Induced Single-Event Burnout in SiC MOSFETs," *IEEE Transactions on Nuclear Science*, Vol. 61, No. 6, 2014, pp. 3109–3114.

[25] Asai, H., et al., "Terrestrial Neutron-Induced Single-Event Burnout in SiC power diodes," *IEEE Transactions on Nuclear Science*, Vol. 59, No. 4, 2012, pp. 880–885.

[26] Griffoni, A., et al., "Neutron-Induced Failure in Silicon IGBTs, Silicon Super-Junction and SiC MOSFETs," *IEEE Transactions on Nuclear Science*, Vol. 59, No. 4, 2012, pp. 866–871.

相关图书推荐

《功率半导体器件——原理、特性和可靠性(原书第2版)》

Semiconductor Power Devices: Physics, Characteristics, Reliability, 2nd edition

约瑟夫·卢茨(Josef Lutz)
[德] 乌维·朔伊尔曼(Uwe Scheuermann) 著　　卞抗　杨莺　刘静　蒋荣舟　等译
里克·德·当克尔(Rik De Doncker)

内容简介：

本书原作者长期从事功率半导体器件的研究和教学工作，在封装、可靠性和系统集成方面做出了重要贡献，在国际上享有盛誉。本书是一本精心编著、并根据作者多年教学经验和工程实践不断补充更新的经典图书，对于广大的研制和生产各种各样的电力电子器件的工程技术人员是极其宝贵的。内容包括：

- 讲述了功率半导体器件的原理、结构、特性和可靠性技术；
- 重点介绍了MOSFET、IGBT等现代功率器件，以及近年来有关功率半导体器件的最新成果，如SiC、GaN器件，以及场控宽禁带器件等；
- 重点阐述了功率半导体器件的制造工艺、测试技术和损坏机理分析。

《SiC/GaN功率半导体封装和可靠性评估技术》

[日] 菅沼克昭　等著
何钧　许恒宇　译

内容简介：

- 本书重点介绍宽禁带功率半导体封装的基本原理和可靠性。
- 以封装为核心，内容涵盖宽禁带功率半导体的模块结构和可靠性问题，引线键合技术，管芯背焊技术，模制树脂技术，绝缘基板技术，冷却散热技术，可靠性评估和检查技术等。
- 尽管极端环境中的材料退化机制尚未明晰，书中还是总结设计了新的封装材料和结构设计，以尽量阐明未来的发展方向。
- 本书对于我国宽禁带(国内也称为第三代)半导体产业的发展有积极意义，适合相关的器件设计、工艺设备、应用、产业规划和投资领域人士阅读。